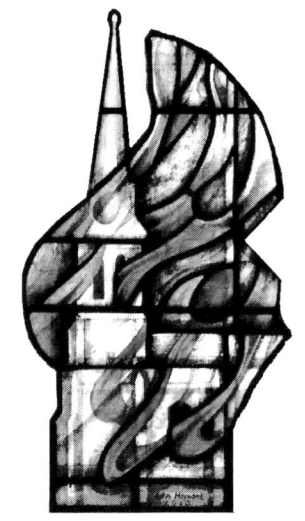

ST. LEONARD'S STREATHAM

A guide to an ancient parish church
that rose from the ashes
after a fire in 1975

by

John W Brown

LOCAL HISTORY PUBLICATIONS

316 GREEN LANE, STREATHAM, LONDON SW16 3AS

Published by
Local History Publications
316 Green Lane, Streatham, London SW16 3AS

Copyright ©2015 John W Brown
ISBN 978 1 910722 01 5

All rights reserved. No part of this publication may be reproduced, stored in a retrieval system, or transmitted, in any form, or by any means, electronic, mechanical, photocopying, recording or otherwise, without the prior permission of the publisher and copyright holders.

ACKNOWLEDGEMENTS

My thanks and appreciation to the following for the use of their photographs and illustrations or for their help and assistance in the preparation of this book: Brian Bloice, Colin Crocker, Croydon Local Studies Library and Archives Service, Charles Gibson, Ken Gordon, Graham and Marion Gower, Andrew Hadden, Mimi Heinrichs, Andrew Hicks, Revd. Mandy Hodgson, Keith Holdaway, Christine Jones, Dr. Margaret Kesterton, John Kinsman, Lambeth Archives, Ann Morisy, Tony Nunn, John Quickenden, Jenny Shaw, The Streatham Society, Janet Weeks, Laurie Whiting and the Revd. Jeffry Wilcox.

IN MEMORY OF GEOFFREY PAGE

This book is written in memory of my good friend and neighbour, Geoffrey Page. He was churchwarden of All Saints church in Sunnyhill Road, Streatham, which was a chapel of ease to St. Leonard's church. The two churchwardens of All Saints were deputy churchwardens at St. Leonard's. He was also PCC secretary and a sidesman at St. Leonard's. On June 13th 1944, a week after D Day, the Archbishop of Canterbury visited St. Leonard's church and in the official picture that was taken to commemorate the visit Geoffrey can be seen standing behind the Archbishop holding the processional cross. He also attended the church of St. Mary, Aldermary, in the City of London, where he was churchwarden with the famous poet, Sir John Betjeman, and both men became friends through this association. Geoffrey died on 28th December 1980. He was a true gentle-man, a compassionate Christian and a wonderful friend.

DEDICATION

This book is dedicated to the Revd. Joshua Rey, his wife Annie and their son Paddy. Joshua has been curate at St. Leonard's church for the past three years during which time he has enriched every aspect of the life of the church. This dedication carries with it the heartfelt thanks and good wishes of the parish for his greatly appreciated contribution to our worship and activities and for the role he played in helping to secure funding of almost a quarter of a million pounds for the renovation and repair of the north slope of the church roof and the buttresses of the church tower through the generosity of English Heritage, Viridor Credits Environmental Company Ltd., National Churches Trust, Garfield Weston Foundation, Allchurches Trust Limited and the congregation of St. Leonard's.

ST. LEONARD'S CHURCH - STREATHAM

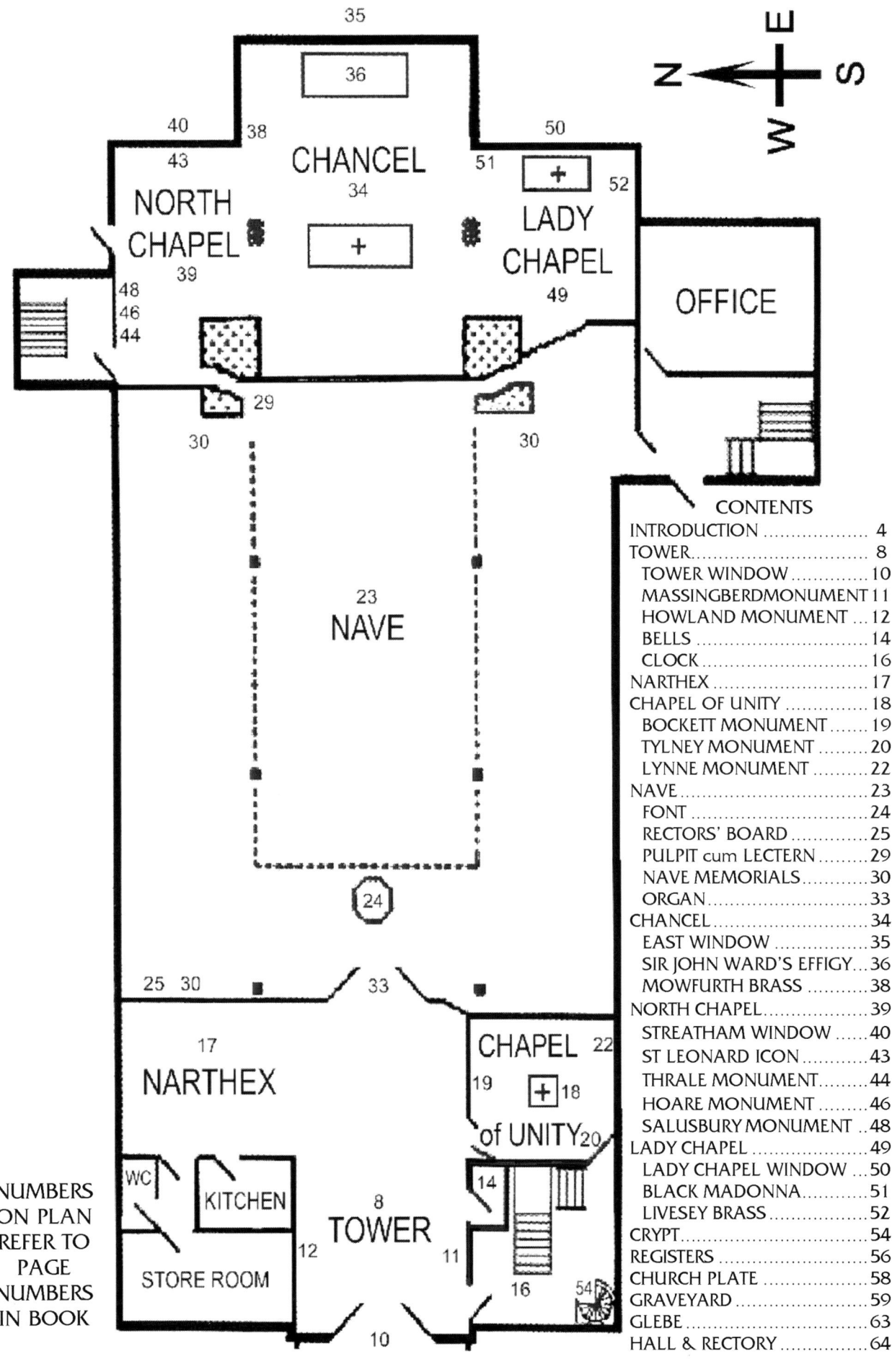

NUMBERS ON PLAN REFER TO PAGE NUMBERS IN BOOK

CONTENTS
INTRODUCTION	4
TOWER	8
TOWER WINDOW	10
MASSINGBERD MONUMENT	11
HOWLAND MONUMENT	12
BELLS	14
CLOCK	16
NARTHEX	17
CHAPEL OF UNITY	18
BOCKETT MONUMENT	19
TYLNEY MONUMENT	20
LYNNE MONUMENT	22
NAVE	23
FONT	24
RECTORS' BOARD	25
PULPIT cum LECTERN	29
NAVE MEMORIALS	30
ORGAN	33
CHANCEL	34
EAST WINDOW	35
SIR JOHN WARD'S EFFIGY	36
MOWFURTH BRASS	38
NORTH CHAPEL	39
STREATHAM WINDOW	40
ST LEONARD ICON	43
THRALE MONUMENT	44
HOARE MONUMENT	46
SALUSBURY MONUMENT	48
LADY CHAPEL	49
LADY CHAPEL WINDOW	50
BLACK MADONNA	51
LIVESEY BRASS	52
CRYPT	54
REGISTERS	56
CHURCH PLATE	58
GRAVEYARD	59
GLEBE	63
HALL & RECTORY	64

INTRODUCTION

St. Leonard's church is the oldest building in Streatham with the tower dating back to the 1350s. Indeed, it is claimed to be the oldest structure on the A23 London to Brighton road between Lambeth Palace and Croydon Palace.

A small Roman votive figure discovered when digging the foundations of the Roman Catholic Church of the English Martyrs suggests the area around St. Leonard's may have originally been the site of a pagan shrine in Roman times.

We know there has been a place of Christian worship here since at least Saxon times and a small chapel, valued at 8 shillings (40p), is recorded here in the Domesday survey of 1086.

The church is dedicated to St. Leonard, the patron saint of the Abbot of Bec in Normandy, on whose Abbey the manors of Streatham and Tooting Bec were endowed by Richard fitz Gilbert, Lord of Clare, who had acquired them by gift of his cousin, William the Conqueror, after the Norman conquest in 1066.

The oldest surviving part of the church is the flint tower which dates from the mid-14th century, when the church is believed to have been rebuilt by Sir John Ward, whose mutilated effigy rests beneath the east window of the church.

St. Leonard's has been rebuilt on at least three subsequent occasions, the last time being in 1975 following a fire which gutted the building and left only the outer walls and tower standing.

St. Leonard's church in the 1600s

St. Leonard's church in the 1790s

St. Leonard's in the mid-1820s before the rebuilding

St. Leonard's church in 1845 after the rebuilding of 1831

The present-day body of the church dates from a rebuilding in 1831. Many of the ancient monuments and fittings of the old church were incorporated into the new 1831 church and some of these have survived down to the present day.

Unfortunately the rebuilding of the church in 1831 was far from satisfactory. The whole process appears to have been fraught with problems and while the church was being built some parishioners were alarmed at the quality of the work being undertaken. Chief among these was Mr. Benjamin Harwood who considered the workmanship to be of a very poor standard, with the drains and foundations insufficient for their purpose. He published several pamphlets on the subject; agitated the vestry to hold meetings to discuss the matter and pressed for local builders and tradesmen to be employed on the project. Many of his criticisms were later found to be well justified; the quality of the work was very poor and the new church had to be closed as it was considered unsafe.

The Bishop of Winchester wrote 'I cannot take upon myself the responsibility of recommending that Divine Service be performed there again at present'.

The church sued the architect, Joseph Parkinson, and builders, Messrs Haynes, Skinner and Borsley, for their shoddy work and the arbitrator awarded St. Leonard's damages against the architect of £200 and £375 against the builders. The legal costs incurred also had to be met by Parkinson (£497 10s) and the builders (£463 10s).

For around 13 years the roof of the church had to be supported by scaffolding poles and it was not until 1844 that Thomas Cubitt, the Clapham Park builder, was engaged to rectify matters.

As part of the rebuilding in 1831 a crypt was formed under the church, the sale of tombs in which helped defray part of the cost of the building work.

It was subsequently discovered that the pillars in the crypt, which supported the church building above, were hollow, whereas they should have been solid. However, it was considered unsafe to break them open to rectify this and only a few tombs in the north wall of the crypt were bricked in to provide additional support for the wall above. This later led to the false belief that these bricked-up tombs were entrances to ancient tunnels under the church.

Fortunately Cubitt gave good advice and his remedial works were successful and for the past 184 years the church has stood on its existing foundations without any serious problems.

Further enlargement of the church was undertaken in 1863 when the chancel was added. Over the following century minor additions and alterations took place but the nave remained substantially the same as it was in Georgian times.

All this was to change on the evening of Monday 5th May 1975. Early that evening the fire brigade was called out to extinguish a bonfire on the glebe.

However, it now seems likely some of the embers from this fire floated up to the roof of the church where they later ignited the timbers causing the conflagration which was spotted by a resident of St. Mary's Convent, opposite the church, later that evening.

The fire brigade was summoned for a second time but unfortunately by the time they arrived the fire had taken hold and flames were leaping from the church high into the night sky. Despite firemen fighting the blaze well into the early hours of the morning all that was left standing the next day was the tower and outer walls of the building.

Revd. Michael Hamilton Sharp standing among the ruins of the church the day after the fire

Firemen survey the damage the morning after the fire

Communion service held in the graveyard the day after the fire

That morning, as smoke still hung over the remains of the gutted church, the Rector, the Revd. Michael Hamilton Sharp, held a communion service in the graveyard using one of the tombs as an altar.

He immediately declared that the church would be rebuilt and through his enthusiasm, dedication and hard work, strongly supported by the congregation and members of the parish, a new era began for St. Leonard's.

A new church was built to the designs of local architects the Douglas Feast Partnership. Douglas lived at 11a Streatham Common South and was a member of St. Leonard's church. He, his associate, David Roberts, and the Rector shared a vision for the new church which rose like a phoenix from the ashes of the old building and enabled the work of St. Leonard's to continue on into a new millennium.

Most of the day to day decisions concerning the rebuilding were taken by a small committee consisting of the Rector and the two churchwardens, Charles Gibson and Andrew Hicks.

Douglas Feast, ably assisted by David Roberts, assembled an expert team to undertake the rebuilding work involved comprising Sawyer and Fisher of Epsom as Quantity Surveyors, headed by their partner George Neville; Brian Colquhoun and Partners were appointed Consulting Engineers with senior partner Arthur Lance, who also lived in Streatham Common South for many years, undertaking most of the work; and Dove Brothers (who had built the chancel extension a century earlier) being appointed builders. Sir Bernard Feilden, Surveyor to St. Paul's and

ST. LEONARD'S CHURCH - STREATHAM

Norwich cathedrals, was appointed Consultant Architect to provide Douglas Feast with expert advice on ecclesiastical matters as the need arose and he visited the site three times during construction.

There is one other person whom I must mention in connection with the restoration of St. Leonard's church who is often overlooked: Inger Norholt. This remarkable Danish lady in her 80s was responsible for the reconstruction, repair and renovation of the font, monuments, and carved stonework of the church.

Many of the damaged monuments were taken to St. Andrew's church in Guildersfield Road in 30 large wooden crates where they were laid out on a scaffolding platform above the pews in the side aisle. There, in the dark and cold surroundings of the church, Inger painstakingly stuck all the pieces together under the most difficult of circumstances. What entered St. Andrew's as a pile of rubble left as the wonderfully restored monuments we see today.

I had the pleasure of meeting her at that time, surrounded by smashed stone and marble, and marvelled at the outstanding skill and craftsmanship she exhibited in restoring these ancient relics from St. Leonard's church.

The rebuilt church was re-dedicated by the Bishop of Southwark on the 22nd December 1976 and by May of the following year regular services were once again being held in the building.

In 1999 the final phase of the rebuilding project was completed when the churchyard was landscaped and railings placed around its perimeter.

My memory of St. Leonard's before the fire is that it was a somewhat dark place. The bleak wooden pews, the walls covered from top to bottom in ancient monuments and the rood screen across the chancel all added to the sense of darkness and enclosure.

Inger Norholt restoring the damaged monuments of St. Leonard's church at St. Andrew's church in south Streatham

The interior of St. Leonard's church before and after the fire

What a difference to the St. Leonard's of today with its bright, white surroundings and sense of space and light. The fire was unquestionably a major disaster but the rebuilding was an opportunity which was seized upon to create a new church blending the best of the old with the new.

Nowhere is this more apparent than in the chancel which was originally designed by William Dyce in 1863. This now provides St. Leonard's with a marvellous cathedral-like area of apparent antiquity, much enhanced by the inspired decision to place the 14th century effigy of Sir John Ward beneath a canopy of similar date at the west end of the church.

This book provides a guide to the church we see today which rose from the ashes of the fire that gutted the building on the evening of the 5th May 1975. Each part of St. Leonard's is examined in sequence and today's church is compared with the church as it was before the conflagration.

I personally think one of the best ways to experience the true splendour of St. Leonard's is to attend a choral evensong on a summer's evening, when the light streams in through the stained glass windows and the building echoes to the magnificent sound of one of the finest choirs and organs in south London.

I hope you enjoy your tour of Streatham's most historic building and in so doing remember, St. Leonard's is not just an ancient church it is also a living community of people and it is they who help bring the building to life.

John W Brown
St. Leonard's Church Archivist

ST. LEONARD'S CHURCH - STREATHAM

St. Leonard's after the fire which destroyed the church on the evening of 5th May 1975

TOWER

The tower of St. Leonard's church is the oldest surviving structure between Lambeth Palace and Croydon Palace and has been a prominent local landmark for almost seven centuries.

It was built c1350 by Sir John Ward, a friend of the Black Prince, with whom he fought in France. The tower was erected when the church was rebuilt either as a thank offering for Sir John's safe return from the French Wars or for the deliverance of Streatham from the Black Death.

The tower is built of knapped flints with stone dressings and contains a magnificent 14th century archway and a star-shaped lierne-vault. Much of the original flint and stone work, which once formed the exterior walls of the tower, can be seen today inside the church.

The spiral staircase in the tower, leading up to the belfry, has remained little changed over the centuries although the old stone steps that were worn away with almost 700 years of use have now been encased in wood.

Sitting above the tower used to be a large wooden steeple which was clearly visible from as far away as Brixton Hill. It was hit by lightning several times in centuries past and in 1777 lightning damaged it to such an extent it had to be completely rebuilt.

In 1841 it was again hit by lightning and a plaque on the exterior west wall of the tower, just below the belfry, recalls this incident and reads:

St. Leonard's tower and spire in the 1790s

> The late spire of this church
> was destroyed by lightning
> on the morning of the 3rd January 1841
> and the present spire was erected
> during the same year.
>   Rev. Henry Blunt MA   Rector
>   Samuel Jasper Blunt
>   Joseph Hartnell   Churchwardens

The storm that destroyed the spire in 1841 was one of the most violent experienced in London. It lasted about an hour and half and was accompanied by a fall of large hailstones.

Lightning struck the steeple shortly after 7am on Sunday morning. A young boy passing the church saw the steeple ablaze and ran to tell Mr. Street, the parish clerk, who lived in the nearby forge. Street, accompanied by George Sandy, a bell-ringer, gained entry to the church and climbed the spiral staircase to the belfry where they could see the fire raging just beneath the ball of the weather vane. Thinking he could successfully douse the flames Sandy collected a bucket of water and mop, reclimbed the stairs and attempted to extinguish the blaze but with no success. Fire soon engulfed the old oak shingles covering the spire and when the Streatham fire engine arrived its task seemed hopeless. A rider was sent galloping up the high road to fetch the Waterloo Road fire engine and when this arrived, along with three other pumps, the fire was eventually brought under control. Although the spire was completely destroyed the fire fighters successfully prevented it from spreading to the main body of the church.

George Sandy, who had so valiantly fought the blaze with bucket and mop, was born in Streatham on 29th October 1801 and was baptised at St. Leonard's church on 15th November. He was a farm labourer and lived in Leigham Lane, now known as Sunnyhill Road. He died in 1859, aged 57, and was buried in the churchyard on 27th February.

A new spire was erected later in 1841 at a cost of £650. It was built of bricks and covered in Roman cement in keeping with the design of the church which had been rebuilt ten years earlier in 1831.

Old pictures of the church show that a weather vane with a large copper ball, 2½ feet in diameter, used to sit atop the spire and in 1795 12 guineas (£12.60) was paid for regilding the ball, vane, clock face and hands.

On Sunday 4th January 1959, just after evensong, the weather vane and cross that then sat on top of the spire came crashing down onto the porch roof. It was later discovered that machine gun bullet holes in the base of the cross, dating back to the Second World War, had significantly weakened the structure over the years, causing the metal to finally give way.

The vane and cross were replaced with a large metal cross which survived until the fire that gutted the church in 1975. This cross can still be seen inside the church, where it is kept in the clergy vestry.

ST. LEONARD'S CHURCH - STREATHAM

Revd. Hamilton Sharp with the new weather vane

The lierne-vault of the 14th century tower

In May 1976 a new weather vane, finished in gold leaf and designed by Douglas Feast, was placed on top of the spire where it remains today telling passers-by in which direction the wind is blowing.

Set in the floor of the tower is a tombstone recording the death of Walter Powell who died on 8th April 1802, aged 71, and his wife, Sarah, who passed away on 13th November 1831 in her 92nd year. Walter was a wealthy Lombard Street banker in the City of London and lived in Grove House, a large 10 bedroom mansion on Tooting High Street which was surrounded by extensive gardens and grounds. The house was later known as St. Leonard's.

Since 2007 an antenna to boost local mobile phone services in Streatham has been housed in the steeple, with an accompanying telecommunications base station in the crypt.

For many years a small Holm oak tree grew from the masonry of the tower, to the left of the lightning plaque. This was the highest oak tree in Streatham until it was removed during work on the tower in 2014/15 when the buttresses were recovered in Roman cement and the north slope of the roof was repaired at a cost of almost £250,000.

The tower with the Holm oak tree growing to the left of the lightning plaque

The ornate internal metal gates which protect the tower entrance probably date from the 1831 rebuilding and are likely to have been made by the village blacksmith, Thomas Street, whose forge was opposite the church in Streatham High Road and who is buried in the graveyard.

The tower of St. Leonard's church is Streatham's oldest surviving structure and is by far the most important contributor to the town's ancient architectural heritage. It provides a unique link with Streatham's past and, since it was built seven centuries ago, has witnessed the transformation of the small cluster of houses that stood at the junction of Streatham High Road, Mitcham Lane and Tooting Bec Gardens into the bustling south London suburb we see today.

The ornate metal sliding gates to the tower door

Exterior of the ancient tower inside the church

ST. LEONARD'S CHURCH - STREATHAM

TOWER WINDOW

The stained glass window over the tower entrance to St. Leonard's was designed by John Hayward who created all the stained glass in the present church.

It provides a tangible link with the old church, destroyed by fire in May 1975, as it is made partly from salvaged pieces of stained glass from the old east window of St. Leonard's. This was designed by Lawrence Lee and installed in 1951 to replace the window which was destroyed when a bomb fell in the graveyard on 29th October 1940 (see page 35).

The main part of the new west window comprises a large circle depicting, on the left, the arms of the Archbishop of Canterbury and on the right those of the Diocese of Southwark, to which St. Leonard's has belonged since 1905.

Above these are the arms of the other two diocese to which Streatham has belonged; on the left is Rochester, from 1877-1905, and on the right is Winchester, from ancient times until 1877. These arms were originally in similar positions in the old east window.

At the top of the window can be seen a white dove, representing the Holy Spirit, which crowned the central panel of the old east window. At the bottom is the head of St. Benedict who was depicted in the right hand lancet of the old window.

The old east window had three lancets: that on the left featured St. Leonard but his effigy was completely destroyed by the fire; the central panel featured Christ but again this panel was severely damaged by the fire; on the right was St. Benedict whose head remained suitable for reuse in the west window.

St. Benedict (c480-543) founded the great monastic order of Benedictine monks and is the patron saint of Europe and students. In 1034, Herluin, a Norman knight, founded the Benedictine Abbey of Bec, in Normandy, France, to which the Manors of Streatham and Tooting Bec were endowed by Richard fitz Gilbert, Lord of Clare, who received them from William the Conqueror after the Battle of Hastings in 1066. Bec Abbey administered the manors for around 300 years up to the time of King Edward III.

The stained glass which originally filled the tower window was installed in 1871. It was designed by Clayton and Bell and showed the aged Simeon in the centre holding the baby Jesus with Mary and Joseph on either side. It was erected by James Brand as a memorial to his wife, Mary, who died aged 31 and was buried in the family vault in the graveyard on 17th May 1870. Her body was removed for re-interment at Sanderstead on 19th December 1893.

This window was damaged beyond repair during the 1975 fire and was the first window to be replaced when the church was rebuilt.

Centre: Abbey of Bec on a French stamp issued in 1978
Bottom: Remains of the ancient Benedictine Abbey of Bec in the late 1700s

MASSINGBERD MONUMENT

On the south wall of the tower is a handsome mid-17th century monument to John Massingberd, a wealthy merchant and treasurer of the East India Company.

It was originally on the north wall of the chancel but when the church was rebuilt in 1831 it was re-erected in the tower, opposite the Howland memorial.

The monument shows John and his wife, Cecilia, kneeling either side of a faldstool, a design that was a special feature of the Southwark school of alabaster sculptors in the reigns of Elizabeth I and James I. At the time of the erection of the monument in 1653 the style had gone out of fashion but no doubt it was one that appealed to the Massingberds and depicts them as a devout couple at prayer.

The inscription reads:

Here lyeth the body of
JOHN MASSINGBIRD ESQUIRE
who departed this life XXIII November MDCLIII
leaving CECILIA his wife
with two daughters
ELIZABETH and MARY
the elder married some yeeres before to
GEORGE BERKELEY
only son of the LORD BERKELEY
the younger since to
ROBERT LORD WILLUGHBY
Eldest son of the EARLE OF LYNDSEY

John was buried in a tomb in the old church on 30th November 1653 and his wife joined him there almost two years later on the 3rd August 1655.

To the left of the inscription is the Massingberd coat of arms. Note the boar with a curly tail with a red cross on its flank which refers to Sir Oswald Massingberd (c1522) who was a Knight of St. John of Jerusalem for which the red cross was their emblem.

The family motto is 'Est meruisti satis' (It is enough to have been worthy).

The Massingberds lived in a large house called Broadwaters, which stood on an ancient moated site which had previously been the manor house for Tooting Bec. The site is now occupied by Knapdale at 21 Tooting Bec Road.

John was one of a number of wealthy residents of Streatham with connections with the East India Company.

One of the company's vessels was named the Massingberd and John occupied the important position of treasurer to the company for nine years.

John's commercial activities brought him much wealth and a failed prosecution for exporting gold and silver without having permission in 1637 suggests he may have been an adventurous trader.

This was not John's only brush with officialdom for in 1641 he was fined £1 by the Lord of the Manor for digging up sand in the King's Highway to the nuisance of the local villagers.

The monument tells us that his two daughters both married well. In 1646 the eldest, Elizabeth, married George Berkeley, the only son of Lord Berkeley. In 1660 George was a member of the Commission to go to the Hague to invite King Charles II to return to England to take up the throne. In 1665 he went to France to look after the King's mother in Paris.

He was made an Earl in 1679 and went on to be a member of the Privy Council and became Master of Trinity House. He was also a member of the temporary government that served after the fall of King James II.

George was a founder member of the Royal African Company and was an active member of the Royal Society. He was also a member of the East India Company. He died in 1698 and his wife, Elizabeth, passed away ten years later in 1708.

John Massingberd's younger daughter, Mary, married Robert Lord Willoughby who later became the 3rd Earl of Lyndsey.

HOWLAND MONUMENT

On the north wall of the tower is the largest and grandest monument in St. Leonard's church which was erected in 1686 in memory of John Howland. The memorial is of the highest quality as is evident in the carving of the drapery, the flower wreaths, the palm branches, cherubs etc.

The monument is probably by the great 17th century sculptor, John Nost. It incorporates many symbolic features with cherubs weeping over an urn and a laurel-crowned skull representing mortality; and lighted lamps, a flowing urn and flower wreaths to represent immortality.

John Howland was Lord of the Manor of Streatham and Tooting Bec. The estate came into the possession of the Howland family in 1599 when it was purchased for £3,000 by Giles Howland, a member of the Grocers' Company and a founder subscriber of the East India Company. He was knighted in 1603. His second wife was Elizabeth, the daughter of Sir John Rivers, Lord Mayor of London in 1573. In his will Sir Giles bequeathed the sum of £5 per annum in perpetuity to be distributed in bread to the poor of the parish of Streatham.

On Sir Giles' death, in 1608, his son, John, inherited the manor and in 1610 he greatly added to his local land holdings with the purchase of Leigham Court Manor. He was knighted in 1616. Sir John died in 1649 and was buried in St. Leonard's church on November 7th. As well as leaving £10 in his will for a new pulpit for St. Leonard's church he bequeathed £5 a year in perpetuity to augment his father's bread dole so that 20 poor people of the parish would receive a loaf of bread every Sunday. Up until the fire in 1975 the bread dole box, with its 20 pigeon holes to accommodate the loaves, could still be seen at the west end of the north aisle of the church. The bread dole was discontinued in 1886 and the money is now administered by the Streatham Charity Trustees. Sir John also left £1 6s 8d (£1.66) per annum to be paid for the preaching of a sermon on Christmas Day at St. Leonard's church.

In 1648, shortly before his death, Sir John sold the Manor of Streatham and Tooting Bec to his brother, Geoffrey, who was buried in the family vault in the chancel of the church on Christmas Eve 1679. It was Geoffrey's son, John, who inherited the manor and in whose memory the monument was erected.

John married Elizabeth, the daughter of Sir Josiah Child (1630-1699), chairman of the East India Company. Sir Josiah was an extremely successful and wealthy merchant and was said to be worth £200,000. His handsome tomb at Wanstead is also by John Nost.

Elizabeth gave birth to her first child, a daughter, who was christened Elizabeth on the 4th August 1682. Her second child, a son called John, was baptised on 20th March 1684 but died a few months later on the 13th May 1684. Further tragedy hit the family two years later when John Howland died on 2nd September 1686 leaving his wife Elizabeth to administer his vast business and manorial responsibilities.

Elizabeth proved well able for this task and was an astute and diligent businesswoman, managing the family interests in the East India Company and developing the Howland Great Wet Dock at Rotherhithe in association with Lady Rachel Russell, the widow of Lord William Russell. He was the third son of the 5th Earl of Bedford who was created Duke of Bedford in 1694. Lord William was executed in 1683 for his complicity in the Rye House plot to ambush King Charles II near Rye House in Hoddesdon.

Elizabeth and John Howland

ST. LEONARD'S CHURCH - STREATHAM

The Howland bread dole box which used to be in the north-western corner of the church

These two astute widows decided to combine their business interests and secure the future of their children through marriage despite their offsprings' young ages. So it was that on 23rd May 1695 in the chapel of Streatham Manor House the 13-year-old Elizabeth Howland was married to the 14-year-old Wriothesley Russell, Marquess of Tavistock, by the Bishop of Sarum (see page 56).

It is said that after the wedding banquet the young couple could not be found as they had sneaked away to play together in the grounds. When members of the party came looking for them the young Elizabeth was found hiding in a barn, her wedding dress torn to pieces!

As a consequence of the marriage on 13th June 1695 King William III conferred on the Marquess of Tavistock, and his heirs, the title of Baron Howland of Streatham with a right to a seat in the House of Lords. This title continues to be held today by the eldest son of the Duke of Bedford.

After the wedding Elizabeth Howland managed the affairs of her daughter, who remained under her care at their home at Streatham Manor House. Her daughter only met her husband occasionally and did not get to live with him until around 1700 when he completed his education. They then spent much of their time at Streatham.

The young couple became the 2nd Duke and Duchess of Bedford on the 7th September 1700 on the death of the 1st Duke, the Marquess's grandfather. It was in the Manor House at Streatham that the 3rd and 4th Dukes of Bedford were born in 1708 and 1710, both of whom were baptised in St. Leonard's church.

Elizabeth Howland died at Streatham on 19th April 1719, aged 57, and was interred alongside her husband in the family tomb in St. Leonard's church on 1st May. Space seems to have been left on the cartouche for her inscription, which for some reason was never added.

In her will Elizabeth left a number of bequests to the parish including £2 per annum to be paid in perpetuity for the preaching of a sermon in St. Leonard's church on the 17th November each year (the date of the accession to the throne of Queen Elizabeth I) to commemorate the monarch's re-establishment of the Protestant faith in England.

The Marquess of Tavistock and Elizabeth Howland who later became the 2nd Duke and Duchess of Bedford

She also left an annual payment of £3 3s (£3.15) to be distributed on St. Thomas' Day (21st December) to 30 poor widows of the parish.

Elizabeth Howland's charity school for girls in Mitcham Lane in 1840

Her largest bequest was the sum of £22 annually to be paid for the education, and the purchase of clothing and religious books, for 10 poor girls of the parish. She stipulated that £8 of this sum was to pay for 'some poor ancient widow resident in the parish of Streatham of a sober life and that can read well' to teach the girls. This was the first charity school to be established in Streatham. Its work continues today through the St. Leonard's Church of England school in Mitcham Lane which, in 1838, incorporated the Girls' school founded by Elizabeth Howland.

13

ST. LEONARD'S CHURCH - STREATHAM

THE BELLS

The bells of St. Leonard's have rung over Streatham since at least the middle of the 14th century when the existing bell tower was built.

Within the tower, the ancient spiral staircase leading to the belfry is little changed over the centuries although the old stone steps that were worn away with almost 700 years of use have now been encased in wood.

The earliest mention of bells at St. Leonard's occurs in 1547, the first year of the reign of King Edward IV, when an inventory of items in the church includes '3 bells in the steeple'.

Later, in the 16th century, this was increased to five bells, an event commemorated in the name of an old local tavern called The Five Bells which was situated opposite Streatham Green.

In 1704 Elizabeth Howland paid 10s (50p) 'to the Ringers at Streatham' for a peal to be rung on 5th November to commemorate the anniversary of the successful quashing of Guy Fawkes' Gunpowder plot in 1605.

The five bells were recast, at a date unknown, by Lester, Pack and Chapman and weighed 23cwt 2qtr 0lb. This may have been in 1749 when the vestry agreed that John Durnford be excused serving as Churchwarden on his giving 5 guineas (£5.25) towards the cost of casting a new tenor bell.

Concealed behind the modern plaster in the ringing chamber are generations of graffiti left by past bell ringers including the name of Edward Denman and the date 1774. His family ran The Five Bells tavern in Streatham between 1759 and 1768.

Account for ringing the bells to celebrate Nelson's victory at the Battle of Trafalgar in 1805

The spiral staircase to the belfry dating from the 1350s

In 1785 the bells were recast, in the key of F, by William Mears at their foundry at Whitechapel at a cost of £209 11s 8d (£209.58). The Duke of Bedford, Lord of the Manor of Streatham and Tooting Bec, donated a 6th bell, a tenor, at a cost of £99 12s 6d (£99.62½) on condition it be the clock bell on which the hour was struck. It weighed 12cwt 3qtr 21lb and had a diameter of 42 inches.

On November 6th 1805 the bell ringers were paid 6s (30p) for ringing a celebratory peal 'Rejoicing at Nelson's Victory' at the battle of Trafalgar and in 1820 the bell ringers were provided with four gallons of beer at 6s 8d (34p) and bread and cheese worth 1s (5p) to sustain them during the ringing to celebrate the coronation of King George IV.

At this time the ringers sought any excuse to practise their art, so much so that in 1829 the vestry ordered 'the ringers be restrained from ringing except on New Year's Day, the Queen's birthday, the King's birthday, King Charles' Restoration, the Prince of Wales' birthday, the King Crowned and gun-powder treason' unless ordered otherwise.

In 1906 John Warner & Co. recast the Tenor bell and made two new bells for the tower, one in memory of Canon Revd. John R Nicholl (1809-1905), the longest serving Rector of Streatham, who was incumbent at St. Leonard's for 61 years from 1843-1904. This peal of eight bells was destroyed in the fire which gutted the church in 1975.

The ringing chamber in the 1930s

ST. LEONARD'S CHURCH - STREATHAM

The present ring of eight bells, in the key of G, were cast at the Whitechapel Bell Foundry in London using 45cwt of bell metal salvaged from the old bells. The new bells were cast in 1980/81 and these were hung by volunteers and dedicated on Sunday 25th October 1981.

This ring comprises:

Treble (G) 3cwt 3qtr 0lb - Named **JOHN 1960-77** in memory of John Carrie, a ringer at St. Leonard's, who was tragically killed in a motor cycle accident at the age of 17.

2nd (F#) 3cwt 3qtr 16lb - Named **+MERVYN SOUTHWARK 1959-80** after the late Rt. Revd. Dr. Mervyn Stockwood, Bishop of Southwark, at the time the bells were recast.

3rd (E) 4cwt 0qtr 16lb - Named **ELIZABETH** to commemorate the recasting of the bells during the reign of Queen Elizabeth II.

4th (D) 5cwt 0qtr 8lb - Named **BEDFORD** to commemorate the gift of the old tenor bell in 1785 by the Duke of Bedford, the former Lord of the Manor of Streatham and Tooting Bec.

5th (C) 5cwt 3qtr 20lb - Named **LEIGHAM** after the Manor of Leigham in which a major part of Streatham was situated and dedicated to the people of Streatham.

6th (B) 6cwt 3qtr 25lb - Named **DOUGLAS** after Douglas Feast, the architect who rebuilt the church after the fire.

7th (A) 9cwt 1qtr 10lb - Named **MICHAEL** after the Revd. Michael Hamilton Sharp, Rector of St. Leonard's when the bells were restored.

Tenor (G) 12cwt 3qtr 6lb - Named **LEONARD** and inscribed **THE SURREY ASSOCIATION OF CHURCH BELL RINGERS CENTENARY BELL 1880-1980** By tradition the tenor is named after the patron saint of the church.

St. Leonard's Bell-ringers in 1938

The renovated ringing chamber in 2010

Graffiti left in the belfry by Edward Denman in 1774

Andrew Johnson helping to rehang the new bells in 1981

This bell was the gift of the Surrey Association of Church Bell Ringers, who also provided their expertise and assistance with the rehanging to commemorate the centenary of the association. This was founded in 1880 by bell ringers from Streatham and Beddington and now covers an area stretching from Southwark into rural Surrey beyond Dorking and Redhill.

By 2010 the internal mortar of the ringing chamber, which had been seriously weakened by the fire in 1975, had badly deteriorated and the vibration of the bells caused small pieces of masonry to fall on the ringers making the tower unsafe to use. The church embarked on a £30,000 restoration project to make the ringing chamber safe, as well as to renovate the ancient clock housed in the tower, and this project was completed on 3rd October 2010.

Today, an enthusiastic group of bell-ringers continue to practise their art at St. Leonard's, as their predecessors had done for centuries before them.

As well as ringing each Sunday for services, various special events have brought the bell-ringers to the tower such as on 1st January 2000 when a special peal was rung to mark the commencement of the new Millennium and on 10th December 2005 when the bells tolled to mark the last journey of the 159 Routemaster bus service through Streatham.

On 10th June 2006 a specially written peal of *Jeffry's Delight Major* of 5,024 changes, composed by Tony Nunn, was rung in 2 hours 44 minutes to honour the Revd. Canon Jeffry Wilcox on his retirement after 24 years as Rector of Streatham.

On Sunday November 2nd 2014 the bells of St. Leonard's rang out over the UK when they were featured on the *Bells on Sunday* programme on BBC Radio 4 which was broadcast on the nearest Sunday to St. Leonard's Day (Nov 6th).

ST. LEONARD'S CHURCH - STREATHAM

CLOCK

In days past villagers learnt the time from a sundial fixed high on the wall of the tower facing Streatham High Road. In 1723 a new sundial was purchased for £6 15s 6d (£6.77½) and in 1748 a replacement was bought for £6 6s (£6.30).

We do not know when the first clock was installed at St. Leonard's as the parish records only survive from 1721 but certainly at that time a clock was in use. There is an entry in the parish accounts for that year when £3 15s 0d (£3.75) was paid for 'altering ye clock and mending it'.

The following year 8s (40p) was spent on ropes for the clock, probably used to rehang the heavy weights which powered the mechanism of the clock.

In 1737 John Poplett of Mitcham was paid £4 to mend the clock and received 10s 6d (52½p) a year up to the early 1740s for regular maintenance of the instrument.

In 1746 'Young Parnum' was paid 5s (25p) a quarter for winding the clock and he continued to do so throughout the 1740s, after which it would appear this task was performed on a voluntary basis by others.

In 1779 a new clock was installed which is still in service today, over 236 years later! This instrument was supplied by Charles Penton of Moorfields as recorded on a

The original inscribed clock dial installed by Charles Penton of Moorfields in 1779

The clock mechanism after the fire

The refurbished clock mechanism in 2010

large brass dial attached to the workings.

As the years passed major renovations of the mechanism were required and other brass plates attached to the clock record these as taking place in 1877 and 1934.

Surprisingly the clock was not totally destroyed in the fire in 1975 and had stopped at 10 to 7 on the evening of May 5th, indicating the time when the flames had reached the mechanism. Although badly damaged the clock was restored and brought back to life to enter a third century of service.

In 2010, after over 230 years of timekeeping, the clock needed a major renovation and this was undertaken as part of the Tower Restoration Project when the manual winding of the clock was replaced by an electric motor.

The mechanism of the 1779 clock can be seen by climbing the stairs in the south porch, where it is enclosed in a large glass case suspended on a platform above the landing. Beneath it, the pendulum can be seen gently swinging from side to side.

Today this ancient clock continues to advise Streatham of the time as it has for almost a quarter of a millennium past and long may it continue to do so.

One of the old stone weights that drove the clock mechanism up to 2010

ST. LEONARD'S CHURCH - STREATHAM

Streatham Grammar School in Granville House now the site of the South London Islamic Centre

NARTHEX

The narthex is the entrance lobby at the east end of the church between the tower and the nave. Before the fire in 1975 this area formed part of the nave.

On the floor is a large white tombstone in memory of Elias Durnford Esq. of Norwood who died on 17th May 1774, aged 54. He was Deputy Treasurer of His Majesty's Ordnance at the Tower of London.

There is a stained glass window at the eastern end of the north wall in memory of those former pupils of Streatham Grammar School who gave their lives for their country in the two world wars 1914-18 and 1939-45.

The window shows the school crest either side of which are the dates 1880, the year in which the school was founded and 1961, the year it closed. It was designed by John Hayward and was dedicated at a special service held on 14th March 1981 attended by a large number of old boys of the school.

Streatham Grammar School was founded by Herbert Large at 122 Sunnyhill Road. The school quickly established itself as one of the best boys' schools in the area and moved to larger premises at Granville House at the top of Mitcham Lane, opposite Streatham Green.

In 1898 the school expanded again when a further move took place, this time to a large house called Loanda, at 40 Mitcham Lane, where it remained until the school closed in 1961.

The quality of teaching at the school ensured its continued success between the wars and it attracted pupils from a wide area. Among its former pupils

Streatham Grammar School Memorial Window

are Lord Harris, founder of Carpetright, Lt. Col. Colin Campbell 'Mad Mitch' Mitchell, who commanded the Argyll and Sutherland Highlanders in Aden, Arthur M Skeffington, Labour MP for Hayes and Harlington between 1953-71, the actor Bonar Colleano and the writer Vincent Broome.

The site of Streatham Grammar School at 40 Mitcham Lane is now occupied by St. Leonard's School. In 1968 the junior school moved there, with the infant school completing the move in 1985. The school was extended in 2001 when new offices, a hall and classrooms were opened. It was further enlarged in 2015.

A major reorganisation of the narthex is currently planned which includes a large part of the area being converted into a kitchen, with a serving hatch opening directly into the nave, and the creation of toilets, with disabled access, in the area now occupied by a small kitchen, toilet and storeroom.

Proposed new layout for the Narthex incorporating a new kitchen and toilets

ST. LEONARD'S CHURCH - STREATHAM

CHAPEL OF UNITY

When redesigning the church after the fire it was decided to create a small chapel which could be used for daily services when only a small number of worshippers were present or for quiet contemplation and private prayer when other activities were being held in the church.

It also provided a space where churches of all denominations could worship together. At the time of the rebuilding the Methodist Church were exploring the possibility of sharing the new church and so this space was named the Chapel of Unity. Although this union did not take place it was decided to keep the name and in 1979 the friends of Streatham Methodist Church presented St. Leonard's with the prayer desk which is in daily use in the chapel.

In July 1978, the Bishop of Southwark, Dr. Mervyn Stockwood, consecrated the new stone altar in the chapel which was designed by Douglas Feast. This, together with the semicircular bench on which worshippers sit, was given in memory of Mary Large, and below the window is a small brass plaque which records this gift with the words 'This Chapel was furnished in 1978 to the Glory of God in memory of Mary Large'.

Above the altar, suspended in a hanging pyx designed by Douglas Feast, the sacrament is reserved.

On the floor, next to the altar, is a small brass plate with the Latin inscription *Orate pro anima Domini Johannis Elslefeld, quondam Rectoris istius Ecclie* (Pray for the soul of John Elslefeld, formerly Rector of this church). John was instituted Rector on 5th August 1390. The date of his death is not known as the registers are missing for the period 1415 - 1446.

Praying angel from St. James's church, Piccadilly

Altar designed by Douglas Feast

Crucifix by Jane Quail

This is the oldest surviving brass in the church and formerly was inserted into John's gravestone which was situated to the north of the altar in the old church.

The crucifix on the wall comprises a soft white French stone figure on a slate cross and was presented to the church by Laurie Bailey in memory of his wife, Nina. It was carved by Jane Quail of Norfolk. She was born in India in 1936 and came to Britain in 1960. Her most widely known work is the fifteen Beatitude Stations in the Anglican Shrine gardens at Walsingham.

Below the crucifix is a small stone panel on which is carved the head of an eagle. This formed part of an old memorial which was damaged beyond repair in the fire.

In the south east corner of the chapel are the remains of a small marble arch with an angel with its hands clasped in prayer on the front. This is one of a pair, the other being in the Lady Chapel, which Revd. Michael Hamilton Sharp obtained from St. James's Church, Piccadilly, as an adornment for St. Leonard's.

On the walls of the chapel are three monuments in memory of Harriet and Elizabeth Bockett, Edmund Tylney, and Rebecca Lynne.

Memorial brass to John Elslefeld, Rector of Streatham in 1390

BOCKETT MONUMENT

On the south wall of the chapel is a fine monument to the wives of John Bockett. The first, Harriet, died in 1840, aged 46, and is interred in the crypt. His second wife, Elizabeth Beatrice, died on 5th November 1862, aged 63, and is buried with her husband in the graveyard.

The inscription reminds us how large Streatham parish once was when it stretched to the southern end of Clapham Common. John Bockett moved to Streatham in 1829 and lived in a large house on Nightingale Lane called Westbury. This occupied the site where Clapham South Underground Station now stands.

John was born in 1790 and was a man of independent means. A devout Christian, he was a fervent supporter of the British and Foreign Bible Society (BFBS). He was elected to the Committee in 1834, the Finance Committee in 1848, and became a Trustee in 1852. In 1862 he became Treasurer and served in this capacity until 1869. During this period the Society's new headquarters in Queen Victoria Street were built and his name is on the foundation stone.

John played an active role in parish affairs and from July 1831 until 1870 he was a Guardian of the Poor of Streatham. He was also a member of the church vestry. He was particularly active in the Balham part of the parish and was one of the first two churchwardens of St. Mary's church in 1855. John was also instrumental in the building of St. Mary's church school and donated the land on which it was built. His name is on the stone he laid when the school was opened on the 9th November 1859. He maintained a keen interest in the school, encouraging the pupils and donating money prizes to those who had the best attendance.

In 1862 he resigned as churchwarden following his election as Treasurer of the BFBS. The vestry recorded their thanks 'for his long and valued service' which had been discharged 'with so much earnestness, courtesy and kindness, with so much denial of self.'

John died on 13th May 1871, aged 80, and is buried in the north-eastern corner of the churchyard.

Elizabeth Horley, the organist of St. Mary's church, Balham, published a pamphlet entitled *Reminiscences of the late John Bockett* in which she described him as being 'extremely conscientious and scrupulously self-denying that he might give the more to those who had but little'.

The inscription on the memorial reads:

In memory of
HARRIET
wife of JOHN BOCKETT ESQRE
Born August 28 1794
Died at Clapham Common in this Parish
31st July 1840
interred in a vault beneath this church
'All her sorrows left below
and earth exchanged for heaven'
Also of ELIZABETH BEATRICE
wife of the above
Born 7th Novr 1798
Died at Clapham Common in this Parish
Novr 5th 1862
Interred in a vault in the Churchyard
Revelations 14-13

Inscription on John Bockett's tomb in the graveyard:

JOHN BOCKETT Esq
Died 13th May 1871
Aged 80 years
ELIZABETH BEATRICE BOCKETT
wife of JOHN BOCKETT Esq
Died 5th November 1862
Aged 63 years

St. Mary's Church, Balham, in the late 1800s

ST. LEONARD'S CHURCH - STREATHAM

EDMUND TYLNEY MONUMENT

The Tylney monument in the Chapel of Unity is one of the historic treasures of St. Leonard's.

It was erected in memory of Edmund Tylney, Master of the Revels to Queen Elizabeth I and King James I, who was buried in the church on 6th October 1610.

Although at the time of his death Edmund was living in a large house in Leatherhead, Surrey, called The Mansion, his family home was in Streatham where his parents had lived.

His father, Philip Tylney, was buried here on 10th September 1541. Despite having been the Gentleman-Usher of the Privy-Chamber to King Henry VIII, Philip died in debt. He was the youngest son of Sir Philip Tylney, Knight Baronet and Treasurer of the Scottish Wars under Thomas, Duke of Norfolk, who firstly married Tylney's cousin, Elizabeth, and later took as his second wife, Philip's sister, Agnes.

We know that Edmund was living in Streatham in 1581 as an entry in the parish registers dated 10th July records the burial of John Hilton, 'servant to Mr Tylney'.

An image of Edmund is depicted in the 'Streatham window' in St. Leonard's church. This shows Edmund dressed in his finery and wearing a ruff with a gold chain around his neck, and sporting a handsome beard and moustache.

In 1572 Edmund joined his cousin Charles Howard (later to become Lord Admiral of England) in the House of Commons as burgess for Gatton in Surrey. It was Charles who secured Tylney the post of Master of the Revels in February 1578.

The revels office arranged entertainment at court and Tylney did much to make its operations more efficient. In 1583, he created the Queen's Men, an elite company of actors picked from the finest performers available, who then dominated court theatricals. Part of his duties was to censor plays and theatrical performances and in this capacity he no doubt had dealings with William Shakespeare and other prominent playwrights of the Tudor period.

It has been suggested that Tylney was a vain and ambitious man as indicated by his memorial. It is adorned with various coats of arms, none of which were awarded to him, but were those of his relatives. Indeed, the majority of the text on the monument is taken up by listing all the important people he was related to. It is interesting to note that among the details of his illustrious forebears there is no reference to one of his most regal relatives, Catherine Howard, Queen of England and the 5th wife of Henry VIII. The reason, no doubt, was because Catherine was beheaded by Henry VIII in 1542 for committing adultery with Thomas Culpepper among others, including a Henry Mannox of Streatham.

Not only this, but Edmund's mother, Malyn Tylney, as lady in waiting to the Queen, was implicated in the scandal and in 1543 was sentenced to life imprisonment and loss of goods, although she was subsequently pardoned after the principals were executed.

Tylney had the monument made during his lifetime and mention is made in his will that it be set up in the church in the place agreed between himself and the parson and churchwardens at Streatham within six months of his decease. The monument cost 20 marks - £13 6s 8d (£13.33) and was made by a 'Stone Cutter neare unto Charingcrosse' believed to be William Wright.

In his will Tylney made several bequests relating to Streatham. The then Rector, Michael Rabit, received half of Tylney's books and a 'great silver Bowl with ewer' for no doubt agreeing to have Tylney's monument erected in a prominent place in the church and for overseeing the execution of his will.

In addition Edmund requested that all his fine clothes were to be sold and the proceeds divided between the poor inhabitants of the Parishes of Leatherhead and Streatham. He also bequeathed unto 'thirteen poor old men and women' of Streatham a double issue of bread, and that they each be given a black gown and 5s (25p) apiece.

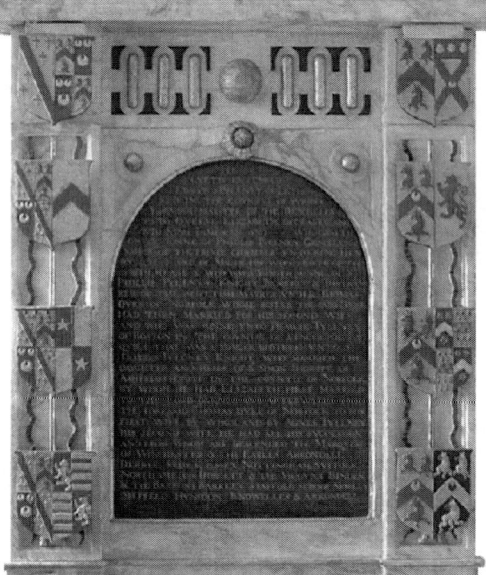

Edmund Tylney as featured in the Streatham Window at St. Leonard's church

ST. LEONARD'S CHURCH - STREATHAM

Tylney is one of Streatham's most important links with Tudor London, the Elizabethan Court and, of course, William Shakespeare. That Edmund held Streatham in high regard is evident by his wish to be buried here, where his mother and father were interred, rather than at Leatherhead where he lived in the largest house in that parish. The bequests he made to the poor of Streatham also indicates a regard for the well-being of the less fortunate inhabitants of our town. But one only has to look at his magnificent memorial to see that it was also his wish that the parish where his family lived should remember him as a man of substance, prestige and importance.

It will be noted the date of Tylney's death on line 6 is missing. Local tradition holds this is because he left no money in his will to have it inserted!

However, the parish registers show he was buried here on 6th October 1610 and is recorded as 'Mr Edmund Tylney Esq Master of the Kings Revils', at which time he had continued in post serving King James I following the death of Queen Elizabeth I.

When St. Leonard's church was rebuilt in 1831 Tylney's magnificent monument was carefully removed. However, when it was re-erected a major error was made in its reconstruction as the top frieze, which is located below the upper coat of arms and above the inscription, should have been placed at the bottom of the monument as is evident from an old drawing of the memorial held in the Guildhall Library in London. This also shows that the marble backplate, behind the coat of arms at the top of the monument, was not re-erected, neither were the three obelisks either side and above the arms. These may have been damaged when removed from the old church or stolen or lost during the rebuilding works.

Today there is a pub called the Edmund Tylney in Leatherhead named in his honour.

In 1998, Tylney was one of the characters featured in the multi-Oscar winning film *Shakespeare in Love*, his part being performed by the actor Simon Callow. It is a fascinating coincidence that both Simon Callow and Edmund Tylney lived in Streatham. Simon was born on 15th June 1949 in a nursing home at 2 Rutford Road and spent his childhood at 4 Pinfold Road with his mother, grandmother and aunt.

The inscription reads as follows:

Drawing of Tylney monument showing correct position of bottom frieze

The Streatham actor Simon Callow playing Edmund Tylney in the film 'Shakespeare In Love'

Heare lyeth interred EDMUND TYLLNEY of Lethered in the County of Surrey Esquire Maister of the Revelles unto QUEENE ELIZABETH Deceased and unto KING JAMES in Ano (*No date inserted*) who was the onely Sonne of PHILIIP TYLLNEY Gentleman usher of the Privie Chamber unto KING HENRY the 8 and of MALIN his wife both of them buried heare & who was a younger sonne of Sir Phillip Tyllney Knight Baronet & Treasurer of the Scottish Warres under THOMAS DUKE OF NORFOLK whose sister the said Duke had then married for his second wife and who was the sonne unto HEWGHE TYLLNEY of Boston in the Countie of Lincolne Esquire that was a younger brother unto SIR FREDRICK TYLLNEY Knight who married the daughter and heire of SIR SIMON THORPPE of Ashfeldthorppe in the County of Norfolk by whom he had ELIZABETH first married to the LORD BARNES and afterwards unto the foresaid THOMAS DUKE OF NORFOLK to his first wife by whom, and by AGNES TYLLNEY his second wife, he had all his succession and from whom are decended the MARQUESS OF WINCHESTER and the EARLS OF ARRONDALL, DERBYE, SUSSEX, ESSEX, NOTTINGHAM, SUFFOLK, NORTHAMPTON, DORCETT & the VICOUNT BINDEN & the Barons BARKLEY, STAFFORD, SCROOP, MORLEY, SHEFFELD, HUNSDON, KNOWELLES & ARRONDALL

ST. LEONARD'S CHURCH - STREATHAM

REBECCA LYNNE MONUMENT

On the north wall of the chapel is a memorial from the Cromwellian period. It is far less flamboyant than the Tylney monument nearby, indicating the stern Protestant ideals of the times.

When restored, after the fire, some of the features were highlighted in gold leaf although it is unlikely that such embellishments would have appeared when the monument was first erected.

Oliver Cromwell

The inscription panel is in black marble with a plain ornamental border surmounted by a heraldic shield, with two more shields at the upper right and left corners. The coats of arms on these shields have been lost in the passing of time.

A skull and crossbones, representing death, are carved in the border either side of the panel and a cherub's head and wings are centrally positioned at the base.

At the top, beneath the heraldic shield, is a heart beneath which are two arms with hands clasped in love.

The memorial was erected by William Lynne in memory of his wife, Rebecca, and the inscription tells us of the great love which they shared.

In Memorie
of my vertueous wife REBECCA LYNNE
Daughter of MR SAMUELL GARRARD
of London Merchant
by whom I had 4 sonnes & 3 daughters
Obiit 2 August 1653
A faithful lovinge wife, more humble deere
was never borne (although borne to lye here)
were Solomon on Earth he would confesse
I found a wife, in whome was happinesse,
SARA, REBECCA, RACHELL all these three
had not more duety wisedome love than shee
with MARY, she did chuse the better part
embracing Christ her saviour, in heart
unto her Mother church a child most true
though of that number, there are now but few
to heaven shees gone, there a place to have
by her redeemer Christ, who his doth save
a vertueous wife on Earths the greatest blessed
O then unhappy I that doe her misse
should I ten thousand yeares enjoy my life
I could not praise enough soe good a wife
WILLIAM LYNNE

We know little about the Lynne family other than what is revealed in the inscription. On the 4th January 1649 'Charles, son of Mr William Lynne' was baptised in St. Leonard's church and there are entries in the burial registers for Rebecca, on the 5th August 1653, and for 'Master William Lynne' on the 3rd March 1659.

The inscription tells us that Rebecca was 'unto her Mother church a child most true though of that number, there are now but few' indicating she remained loyal to the Anglican church through the religious turmoils of the Cromwellian period.

The country was plunged into Civil War three times during the last ten years of Rebecca's life.

In 1644 the Prayer book and Christmas celebrations were banned and in 1649 King Charles I was beheaded.

Four months after Rebecca's death, Cromwell became Lord Protector of England in December 1653.

However, it is the deep love that William had for his wife which shines through this epitaph, summed up exquisitely in the final two lines *'should I ten thousand yeares enjoy my life I could not praise enough soe good a wife'*.

King Charles I

22

ST. LEONARD'S CHURCH - STREATHAM

NAVE

The nave is the main body of the church, between the entrance and the chancel, the area around the altar.

The pre-1831 church was roughly the size of the present nave, contained within the iron pillars which support the gallery and roof, and the south aisle. The north aisle was built on part of the graveyard to provide additional accommodation for the increasing population of Georgian Streatham.

By using iron columns as structural supports much less space was used than by traditional large stone supporting arches. Although iron columns had been used in church building from the 1770s they are rarely seen in south London churches and their use in the rebuilding of St. Leonard's was considered an innovation at the time.

At the base of the steps leading up into the chancel is a large black slab which roughly marks the position of the altar in the pre-1831 church.

To the north of the entrance doors can be seen the old parish chest. Being made of metal it survived the fire.

In the north west corner of the nave, next to the board listing the Rectors of Streatham, is a large wooden cross on which is painted Christ on the Cross. The Revd. Jeffry Wilcox brought this with him from his previous parish in Sunderland when he came to St. Leonard's in 1982. It is a reproduction of the Taize Cross and was made and painted by a parishioner who worked in a shipyard on the River Wear.

Today the floor of the church is covered with flagstones that rest on the roof of the crypt which is made from large, thick, stone slabs. These flagstones accentuate the appearance of antiquity of the church, particularly in the chancel.

Previously wooden pews rested on a timber framework on top of the crypt roof with the aisles covered with ornamental tiles set in several inches of cement.

After the fire it was decided to replace the pews with chairs to enable more flexible seating arrangements to be used. Many of the chairs were given in memory of loved ones and these details are recorded on small brass plaques on each chair.

Left: Wooden Taize Cross made by a shipyard worker in Sunderland
Below: The nave of St. Leonard's church in the early 1900s showing the old pews

ST. LEONARD'S CHURCH - STREATHAM

FONT

The font stands by the entrance to the church. Over the centuries many thousands of baptisms have taken place here. The parish registers date from 1538 and from then until 1900 over 14,000 people are listed as having been baptised at St. Leonard's, the first entry being that for Henry Holland who was baptised on 2nd January 1538.

The carved stone bowl dates from the 15th century and is octagonal in shape with parallel sides decorated with a quartrefoil enclosing a conventional rose which was a common design in the 1400s.

Drawing of 15th century font in 1825

During the fire, which destroyed the church in 1975, the font was smashed in half by timbers and debris falling from the roof. The bowl was restored and if you look closely you can see where the fragments have been pieced together.

Little mention is made of the font in the old church records although there is reference in the church accounts for 1732 of paying 2s 6d (12½p) for 'mending the font'.

In 1894, to commemorate 50 years of service to the parish by the then Rector, the Rev. Canon John R Nicholl, proposals were made for a new font and it is likely it was at this time that a new pedestal was made for the font incorporating eight red marble pillars with stone bases and capitals on which the bowl rested. The font was then raised from the floor to sit at one end of a large stone step which forms the base of the font today.

A wooden octagonal cone-shaped font cover probably also dates from this refurbishment. This survived the fire as it was then no longer in use and was kept in the crypt where it remains today.

After the fire £500 was needed to restore the font and in 1976 a font restoration fund was established. That November the Hilda Holger Contemporary Dance Company performed in the church to raise money for the appeal.

With funding in place it was decided to restore the font to its original 15th century design based on a drawing made in 1825 which shows the bowl resting on a plain octagonal pillar.

The bowl of the font is one of the ancient relics of St. Leonard's where, for over five hundred years, generations of Streatham residents have been baptised into the church.

Font in the early 1900s showing wooden cover

Today's font decorated for Harvest Festival

Font smashed in half during the fire in 1975

ST. LEONARD'S CHURCH - STREATHAM

RECTORS' BOARD

At the west end of the nave, by the south wall, is a large board on which the names of all the known Rectors of Streatham from 1230 to the present day are listed. This was the gift of Maureen Taylor.

Our present Rector, Mandy Hodgson, is the 58th name on the list, and the first woman to hold the post.

Even though there are some minor gaps in the listing, as some of the registers are missing, the board details most of the clergy who have ministered to Streatham during the last 800 years.

Although St. Leonard's church is Saxon in origin, the first recorded Rector was Robert de Rothomago (Rouen) whom we know was in post in 1230 and was probably appointed at an earlier date.

Many of the early Rectors had French names as, up to 1357, they were appointed by the Abbey of Bec in Normandy on which the Manors of Streatham and Tooting Bec were conferred by Richard fitz Gilbert, Lord of Clare, after the Norman conquest in 1066.

The full list of Rectors is as follows

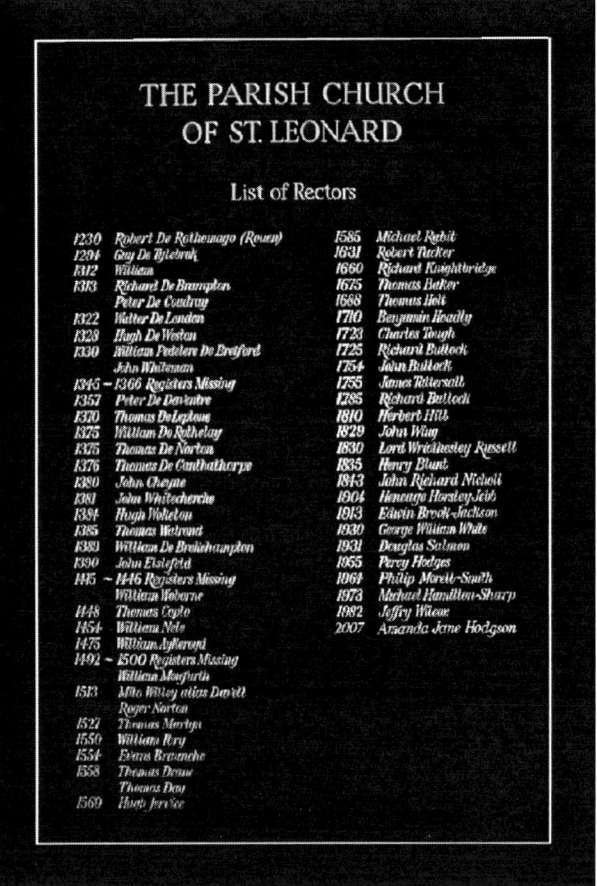

1230   Robert de Rothomago (Rouen)

1294   Guy de Tylebrok

1312   William

1313   Richard de Brampton

1322   Peter de Coudray

1322   Walter de London

1328   Hugh de Weston

1330   William Pedelere de Bretford

Had a dispute with the Abbot of Bec in 1340 concerning the ownership of a house and 60 acres of land in Streatham which was ruled as belonging to the Abbot. This was probably the original Rectory.

-   John Whiteman

Successfully challenged the right of the Abbot of Bec to the ownership of the aforementioned house and land.

(Register 1345-66 missing)

1357   Peter de Daventre

Granted a licence in 1368 to hear confessions, impose penance, and absolve certain crimes excepted in the Archdeaconry of Surrey.

1370   Thomas de Leptone

Appointed by King Edward III after the seizure of French Priories by the Crown due to the war with France.

1375   William de Rothelay

Died in 1375 serving less than a year as Rector.

1375   Thomas de Norton

Resigned in 1375 in exchange for Estandon.

1376   Thomas de Canthathorpe

1380   John Cheyne

Resigned in 1381 in exchange for Trottiscliffe, Kent.

1381   John Whitecherche

Resigned in 1384 in exchange for Stormouth.

1384   Hugh Woketon

Resigned in 1385 in exchange for Aumere.

1385   Thomas Walrond

Resigned in 1389 in exchange for South Wotton.

1389   William de Brokehampton

Resigned in 1390 in exchange for Doeney.

1390   John Elslefeld

His memorial brass is on the floor in the Chapel of Unity (see page 18). He was the Rector in 1394 when the sanctuary of John Brabourne was violated. He was ordered to enforce penance on the three offenders who were to walk in procession in Streatham, stripped to their shirts and drawers, and carrying lighted tapers while Elslefeld flagellated them with a rod (see page 41).

ST. LEONARD'S CHURCH - STREATHAM

(Register 1415-46 missing)

- William Woborne
Resigned in 1448 in exchange for North Wokingdon.

1448 Thomas Copto

1454 William Nele
Resigned in 1475 in exchange for St. Anne's, Aldersgate.

1475 William Aykeroyd
Instituted Rector of Clapham on 4th May 1476.

(Register 1492-1500 missing)

- William Mowfurth
His memorial brass is on the north wall of the chancel (see page 38).

1513 Milo Willey, alias Davell

- Roger Norton
Resigned Aug 3rd 1527 on a pension of £12. Died 26th Dec 1527 and requested he be buried within the chancel of St. Leonard's church under the stained glass window he had installed there (see page 35).

1527 Thomas Martyn
In 1538 Thomas Cromwell ordered that registers of baptisms, marriages and deaths should be kept in every parish. Thomas Martyn promptly obeyed this law so that the registers of our church commence from this date (see page 56).

1550 William Ibry
A Canon of St Paul's Cathedral requested to be buried after the 'old catholic and faithful manner' and left a girdle to each of his nieces on condition that they were catholics.

1554 Evans Braunche

1558 Thomas Deane

- Thomas Day

1569 Hugh Jervice (aka Hugh Gervice)
In his will requested that he be buried 'with Christians in the Chancell at the upper end of Communion table under the great white stone' which was then the high altar at St. Leonard's church.

William Mowfurth

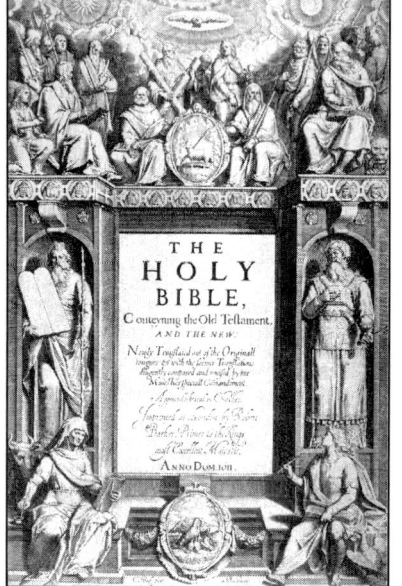
King James Authorised Bible of 1611 on which Michael Rabit worked as a translator.

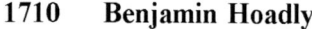
Benjamin Hoadley

1585 Michael Rabit
He was one of the 47 scholars who translated the King James Authorised version of the Bible which was published in 1611. They were organized into six companies, two each working separately at Westminster, Oxford, and Cambridge on sections of the Bible assigned to them. Michael Rabit was one of the Second Westminster Company who were responsible for translating the New Testament Epistles.

1631 Robert Tucker (aka Tooker)
Rector during the time of the English Civil War. In 1657 he was declared to be a lunatic, at which time Richard Knightsbridge was acting as curate-in-charge.

1660 Richard Knightbridge
Started a new parish register on taking up office on the front page of which he celebrated the restoration of the monarchy (see page 56).

1675 Thomas Baker
Meticulous in his keeping of the parish registers; his entries are neat, tidy and clearly written.

1688 Thomas Holt
The grandfather of the celebrated naturalist Gilbert White author of *A Natural History of Selborne*. Holt left £20 a year to the parish which was to be used by the Overseers to relieve the poor until the church rates had been collected. He was present at the marriage of Elizabeth Howland to the Marquess of Tavistock at Streatham Manor House in 1695 (see page 12).

1710 Benjamin Hoadly
Appointed by Elizabeth Howland solely on account of his political views whilst at the same time holding the living of St. Peter le Poor, in Broad Street, London. When King George I became King in 1714 he was appointed his Chaplain and in 1715 he became Bishop of Bangor as a reward for his strong support of the House of Hanover. He became Bishop of Hereford in 1721 and Bishop of Salisbury in 1723 when he

resigned his post as Rector of Streatham. His last promotion was as Bishop of Winchester in 1734 where he lived for 25 years. He died at Chelsea in 1761 at the age of 85. Hoadly Road in Streatham is named in his honour.

1723 Charles Tough
Resigned in exchange for St Paul's Covent Garden. He was elected a Fellow of the Royal Society in 1749 and died 21st June 1754.

1725 Richard Bullock
Also held the living at St. Bride's in London. In 1741 he was fined 1s (5p) for taking turf from Tooting Bec Common for his personal use. Buried at Faulkburne in Essex near his family estate. Father of Richard Bullock appointed Rector of Streatham in 1785.

1754 John Bullock
Died on 17th November 1754 serving less than a year as Rector.

1755 James Tattersall
Rector during the time of Dr. Samuel Johnson's visits to Streatham. He was also Rector of St Paul's Covent Garden 1754-55 and again 1758-84. He built a splendid house called The Shrubbery (opposite Shrubbery Road), which later became Streatham College for Girls (see page 63). He was concerned at the 'great Dissoluteness of regularity and Order that reigns in my Parish'. At the very first vestry he attended on 12th March 1756 he ordered a pair of stocks to be made 'to punnish all offences that be done to the Laws of this Land' and on the 5th December in the same year he approached the Duke of Bedford, the Lord of the Manor of Streatham and Tooting Bec, for land on which to build a cage 'for the confinement of loose and disorderly persons'. This stood by the old smithy opposite the church at the junction of Streatham High Road and Mitcham Lane.

1785 Richard Bullock
Son of Richard Bullock who was appointed Rector of Streatham in 1725. Died 4th October 1809 and was buried in a vault beneath St Paul's, Covent Garden, of which church he was also the Rector.

Robert Southey, Poet Laureate and nephew of Herbert Hill

Henry Blunt

John Richard Nicholl

1810 Herbert Hill
Former Chaplain to the British community in Lisbon, Portugal. Founder of St Leonard's School in 1813. Favourite uncle of Robert Southey, the Poet Laureate, who occasionally stayed at the Rectory with his uncle. Southey is said to have written the fairy story, the Three Bears, for his uncle's children. Hill married Catherine Bigg-Wither, a friend of Jane Austen, the famous author, who visited Streatham Rectory on several occasions and continued to keep in contact with the Hills right up to the time of her death in the summer of 1817 (see page 31).

1829 John Wing
Acted as Curate in charge for Lord Wriothesley Russell between 1830-33 during the period of the rebuilding of St. Leonard's church.

1830 Lord Wriothesley Russell
Brother of the Duke of Bedford, the Lord of the Manor of Streatham and Tooting Bec, in whose gift was the appointment of Rector. Described as a man of considerable attainments he declined the post of Bishop of Lichfield. Rector at the time of the rebuilding of St. Leonard's church in 1831. During his period as Rector many of his Lordship's parochial duties were undertaken by the Revd. John Wing.

1835 Henry Blunt
Former vicar of St. Luke's Church, Chelsea. He was one of the most popular evangelical ministers of his day and attracted large congregations to St. Leonard's who came to hear him preach. Oversaw the building of the parish school in Mitcham Lane and the erection of Christchurch at Streatham Hill.

1843 John Richard Nicholl
Streatham's longest serving incumbent, ministering to the parish for over 60 years from 1843-1904. As a student he attended Eton College where one of his companions was William Gladstone, four times Prime Minister of Britain. When he first came to Streatham it was no more than a small country town. Over the years he watched it steadily grow into a

ST. LEONARD'S CHURCH - STREATHAM

sprawling London suburb of more than 70,000 inhabitants, and raised over £100,000 for the building of 13 new churches in the area to cater for the increased population. Died 10th September 1905 aged 96 years and is buried in the churchyard.

1904 Heneage Horsley Jebb
Descendant of Sir Richard Jebb, physician to the Thrales of Streatham Park (see page 44). Rector at the time of the building of the church hall and the new Rectory, and the rebuilding of the girls' school in Mitcham Lane.

1913 Edwin Brook-Jackson
Resigned to become vicar of St. Mary Abbot's, Kensington. Owned one of the largest private collections in the world of material relating to Napoleon's captivity on St. Helena including several documents signed by the Emperor. He also owned Napoleon's original death certificate. When Brook-Jackson died his vast collection of Napoleonic memorabilia was sold by auction at Sotherby's.

1930 George William White
Curate at St. Leonards from 1916-25. Resigned due to ill health.

1931 Douglas Murray Salmon
Rector during the Second World War. A keen supporter of the Citizens' Advice Bureau which was initially based at the Holy Redeemer church in Streatham Vale before moving to St. Leonard's church hall. Produced the book *To, For and About You the People of Streatham* detailing the activities of various groups in Streatham during the Second World War.

1955 Joseph Percy Hodges
Author of *The Story of St. Leonard's, Streatham* and *The Nature of the Lion - Elizabeth I and our Anglican Heritage*.

1964 Philip Morell-Smith
The last Rector to be appointed by the Duke of Bedford. Former curate of St. Leonard's for 4 years before the Second World War. Resigned to take over the united benefice of Puttenham and Wanborough.

1973 Michael Hamilton Sharp
Rector at the time of the fire which destroyed the church in 1975 and worked tirelessly to ensure its rebuilding.

1982 Jeffry Reed Wilcox
Rector of Streatham for 24 years. Awarded the MBE in the Queen's Birthday Honours in 2005 for services to the local community.

2007 Amanda Hodgson
The first woman to be Rector of Streatham.

The difference between a Rector and a Vicar is that Rectors received their income from the tithes of the parish and were responsible for the upkeep of the chancel of the church, whereas a Vicar received a stipend, or salary, for running the parish. This distinction is now largely historical as tithes have been abolished.

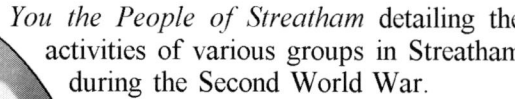
Heneage Horsley Jebb

Douglas Murray Salmon

Michael Hamilton Sharp

Jeffry Reed Wilcox

Amanda Hodgson

ST. LEONARD'S CHURCH - STREATHAM

PULPIT cum LECTERN

The pulpit takes the form of a movable pulpit-cum-lectern and was the gift of St. Leonard's Church School in 1978 in memory of their founder, the Rev Herbert Hill, who was Rector of Streatham between 1810-1828.

It was designed by Douglas Feast and is made from red Canadian pine. Carved on the front of the lectern is a pattern of chains, a reminder that St. Leonard is the patron saint of prisoners.

Prior to the fire St. Leonard's had a richly carved oak pulpit which was a bequest of Sir John Howland, former Lord of the Manor of Streatham and Tooting Bec, who, in 1649, left £10 in his will for a new pulpit for Streatham Church.

Among its fine carvings was the coat of arms of the Howland and Swan families; Cecily Suzan Swan was Sir John's wife.

This pulpit was originally a 'double-decker' with the minister's reading desk at the base, and the pulpit above it surmounted with a sounding board.

In 1774 the pulpit became a 'three-decker' when the structure was raised to allow for a desk to be placed at the base for the use of the parish clerk.

In the early 19th century the pulpit must have presented a magnificent sight as it was then dressed with a deep crimson silk fringe. The comfort of the preacher was well catered for and in 1808 7s (35p) was spent in 'Emptying the feathers out of the Pulpit Cushion, filling it in a new ticken case with materials and making a velvet cover for it'.

In 1831 the pulpit was moved to a central position in the nave of the church. This was not a successful arrangement as when the west door was left open in the summer months it was difficult for the congregation to hear the preacher. To help overcome this problem a wooden screen was fitted under the tower.

When the Rev John Nicholl came to St. Leonard's in 1843 he removed the desks, just leaving the old Jacobean pulpit supported on a wooden column and reached by a small staircase.

On the Sunday following the fire which destroyed the church in 1975, a service was held in the church hall and the cross-shaped iron support, which was all that remained of the Jacobean pulpit, was taken from the ashes and nailed to the wall of the hall, above the altar, to serve as a crucifix.

Left: St. Leonard motif designed by Douglas Feast
Above: The Jacobean pulpit presented to the church by Sir John Howland in 1649
Right: Drawing of the pulpit in 1820 showing the 'sounding board'

NAVE MEMORIALS

Running parallel to the north and south aisles in the nave are rows of tombstones which used to cover the graves of those buried beneath the floor of the pre-1831 church. Some of these memorials have been cropped to make them fit the space available and, sadly, many suffered damage when the roof of the church collapsed to the floor during the 1975 fire.

There are sixteen such memorials in the nave, nine by the north aisle and seven by the south aisle.

HENNIKER FAMILY

By the north aisle, at the west wall, is a stone recording the burial of members of the Henniker family (see page 39).

ELIZABETH LAING

Next east is a stone in memory of Elizabeth Stewart Laing, born 15th May 1752 and died 8th March 1816 aged 63. She was the wife of James Laing of Streatham who was High Sheriff of Surrey in 1815. Sadly, her memorial on the north wall was lost in the fire. The inscription was a heart-felt tribute by her husband which concluded with the words 'This monument, the last tribute of gratitude and affection, is erected by her deeply affected husband, who enjoyed pure happiness in her society during thirty-nine years'. The monument was by Peter Rouw the younger (1770-1852), medallist to the Prince Regent.

James Laing came from Haddo in Scotland. He was originally a doctor and spent many years living in Dominica, where he eventually owned 180 slaves and three plantations, including the Macoucherie Estate which still produces rum under that name today.

James subsequently returned to Britain and spent his retirement in Streatham, living at Hill House, a large property on the northern side of Streatham Common. He resided there from around 1807 until his death, aged 83, in 1831.

In 1812 James presented a clock from his Haddo property to the newly built Crimond church in Aberdeenshire where it was placed in the church tower. It is unique in that it has 61 minutes marked on the clock face with the extra minute occurring between 11 and 12 o'clock, a mistake by the original clock maker. On the clock face are inscribed the words 'The hour's coming'.

James played an active part in Streatham life being a member of several church committees and acting as a local magistrate. He donated £5 towards the cost of establishing the parish school and was chairman of the School Society, presenting pupils with medals won by them in local examinations at a special event held at the White Lion inn on 12th March 1826. He died on 24th April 1831 and his body was interred in the family vault in the crypt of the church.

SOPHIA BORRADAILE

The next stone eastwards records the death on 16th January 1823 aged 11 months and 16 days of Mary Sophia, the daughter of William and Isabella Mary Borradaile who lived in a large house called Gothic Villa on Balham Hill.

THOMAS HOLT

A little further eastwards is the memorial stone to Thomas Holt bearing the family crest. He was Rector of Streatham between 1688 and 1710 and was grandfather of the celebrated naturalist Gilbert White author of *A Natural History of Selborne*. He died on 30th November 1710 aged 56. (See page 26)

Monument to Elizabeth Laing destroyed in the fire showing her husband, James, mourning her loss

Gilbert White grandson of Thomas Holt

Thomas Holt's tombstone

ST. LEONARD'S CHURCH - STREATHAM

JOHN MORGAN

Adjacent is the stone in memory of John Morgan who died on 13th November 1772 aged 32, and his parents, Susanah who died on 12th June 1773 aged 73 and the Revd. Phillip Morgan, Rector of Wasing in Berkshire, who died on 4th June 1774 aged 82.

CICELIA LEE

Next is a large black tombstone dedicated to the 'truely virtuous & religious' Cicelia, wife of George Lee of Lincolns Inn, who died aged 21 on 24th July 1664. She was the grand-daughter of Sir John Rivers of Chafford in Kent (1579-c1651) who was created a baronet in 1621 and was the grandson of Sir John Rivers who was Lord Mayor of London in 1573-4.

JUDITH & FRANCES COVENTRY

The next memorial is to two sisters of the the 5th Earl of Coventry, the Hon. Judith Coventry who died on 27th January 1757, aged 75, and the Hon. Frances Coventry who died aged 74 on 1st September 1762.

MARTHA FIENNES

Abutting the steps is a large black tombstone with a white border in which the inscription has been carved. This stone incorporates a large family crest and was designed by Martha Fiennes, whose death it records on 3rd May 1738 aged 78. She was a wealthy spinster and the sister of Lawrence Fiennes, the 5th Viscount Saye and Sele, to whom she had lent £1,050 at the time of her death. Martha was the daughter of the Hon. John Fiennes and his wife Susannah.

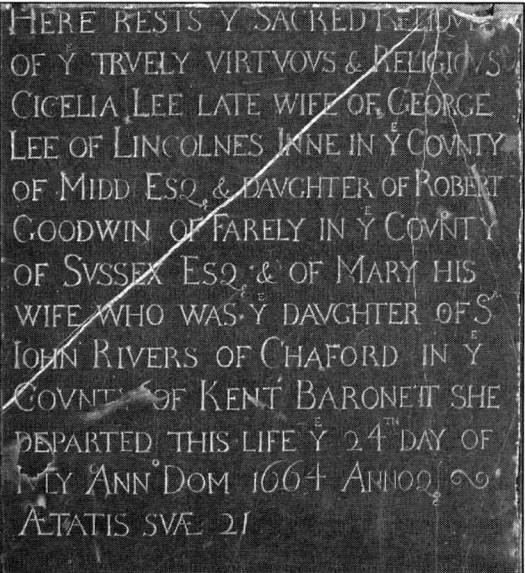
Tombstone of Cicelia Lee

Crest on the tombstone of Martha Fiennes

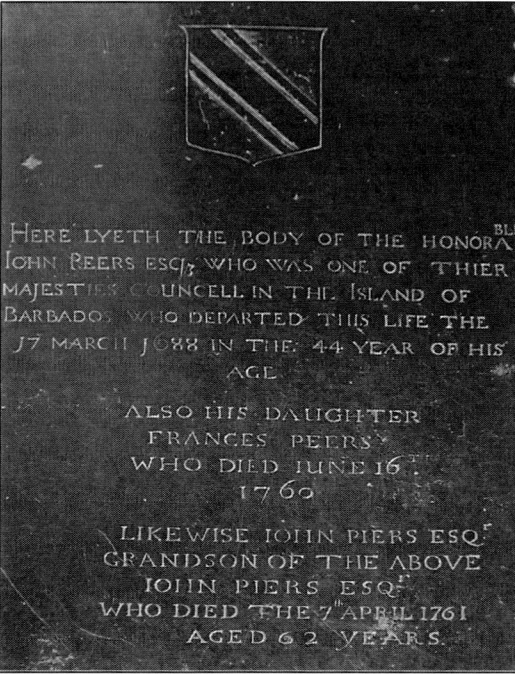
Tombstone of the Hon. John Peers

REVD. HERBERT HILL

To the south of this stone, is the tomb slab of the Revd. Herbert Hill, who died on 19th September 1828, aged 79. It is positioned near the pulpit-cum-lectern presented to the church by St. Leonard's School, which he founded in 1813 (See page 27).

AMELIA STRACHEN

Adjacent to the south aisle, abutting the chancel, there are two memorial slabs side by side. The small white one is in memory of Amelia Strachen who was buried on 29th August 1788. She was the wife of James Strachen, a merchant who traded from 23 Mincing Lane in the City of London, lived in Lower Tooting and died on 7th April 1808. In 1785 he gave £50 to both the Bridewell and Bethlem Hospitals and was subsequently elected a governor of these institutions. In 1788 he was admitted as one of the elder brethren of Trinity House. (The 'T' markings on the memorial show where the base of pews were once set into the stone.)

HON. JOHN PEERS

To the south is the large black memorial stone to the Hon. John Peers Esq. He was born in Barbados where he became the largest plantation owner on the island and a member of the Island Council. He was married twice; firstly c1664 to Hester (aka Elizabeth) by whom he had six children of which 4 survived; secondly in the 1680s to Frances Atkins, the daughter of Sir John Atkins the Governor of Barbados (1674-80). By her he had three daughters, one of whom, Frances, was buried in his tomb. In addition he had three mistresses (two of whom were slave women), each of whom bore him three children all nine of whom were baptised in Christchurch in Barbados on the same day in 1683. John died, aged 43, on 17th March 1688. His daughter, Frances, joined him in the tomb after her death on 16th June 1760 and his grandson, John Peers, followed 10 months later having died on the 7th April 1761, aged 62.

RALPH THRALE & HESTER SALUSBURY

The next stone west is the vault stone for the Thrale family recording the death of Henry's father, Ralph, on the 8th April 1758, aged 60, and his mother in law, Hester Maria Salusbury, on the 18th June 1775, aged 66 (See pages 44 and 48).

HAMBLY FAMILY

Adjacent are two large black memorial stones to members of the Hambly family. The first is in memory of two daughters of Peter Hambly who died in Ashstead; Agnes who died aged 41 and was buried on 24th October 1800 and Mary, who died aged 74 and was buried 23rd November 1819. William Hambly who died aged 82 and was buried on 30th March 1822 is also recorded on the slab.

Ralph Thrale

Abutting is a memorial tombstone with the family crest in memory of William Hambly of Cornwall who died aged 77 on 17th March 1717/8 and his wife, Sarah, who died 13th November 1715, aged 66, who was the daughter of William Hester March. The stone also commemorates the death of William's nephew, Peter Hambly, of Cornwall, a merchant and Citizen of London, who died on 25th May 1723 probably aged 49, although the inscription appears to give his age as 19. In his will he left £5 to the poor of Streatham and £1 10s (£1.50) to the poor of Upper Tooting and Balham.

The Hambly family were prominent residents of Streatham and William moved here from Cornwall in the late 17th century. In 1718 Peter Hambly is recorded as living at Knapdale, a large mansion which stood on the site of 21 Tooting Bec Road, formerly the home of the Massingberd family (see page 11). In 1720 he built a large house opposite Streatham Common, known as Hambly House. By 1722 he owned several properties in Streatham and was granted permission to build new pews in the church to accommodate their residents. Members of the Hambly family continued to live in Streatham until at least 1784 when they moved to Ashstead in Surrey. Hambly House was subsequently acquired by John Henry Lord who established an academy there which became one of Streatham's leading private schools. The house was demolished in 1877 when Hambly Mansions, designed by Ernest George, was erected on the site for Peter Brussey Cow, the owner of the local rubber factory.

SUSANNAH PLATT

The next memorial pays loving tribute to Susannah Platt 'whose tender solicitude for the welfare of her family was as exemplary as her patience and resignation during a long and painful illness which terminated her existence on the 26th of January 1796 in the 59th year of her age'. Her husband, Samuel Platt, is also commemorated. He died on 25th November 1813 aged 84 and was a legal officer under Chief Justice Mansfield. This stone weighs 1½ tons and was smashed into 58 pieces during the fire in 1975 and was restored in 1982.

ELIZABETH KATE WISEMAN

The final tombstone has an impressive family crest at its head and is the oldest stone in the church. It dates from the time of the English Civil War and, although damaged, part of the inscription is still readable although it cannot be collaborated by the burial registers for 1643 as they are incomplete for the period of the Civil War. The stone records the death of Elizabeth Kate, the wife of Mark Wiseman, who died on 19th August 1643.

Hambly House, standing opposite Streatham Common, in the mid-1800s

Crest on Elizabeth Wiseman's tombstone, the earliest in the church dating from 1643

ST. LEONARD'S CHURCH - STREATHAM

ORGAN

At the vestry held on 29th June 1809 Mr Howard proposed 'to erect in the church an organ of the value of £300 at the least, entirely at his own expense, and to play it himself, or provide some proper player for it, trusting to the liberality of the parish for his compensation.' With such a generous offer before them the vestry put the proposal into immediate effect.

By 1828 John Eastman was the church organist at a salary of 6 guineas a year (£6.30), increased to £40 per annum in 1831, and in 1832 John Hubbard was paid £2 a year as 'organ blower' for pumping air into the instrument so it could be played.

After the chancel was extended in 1863 the organ was transferred from the west gallery to the north side of the chancel.

By the early 1900s this organ had seen better days and was 'tied together with bits of wire and string'. The then Rector, the Rev. Canon John R Nicholl, had always resisted calls for the organ to be replaced, claiming it would 'last my time'. Ironically it was played for the last time on 10th September 1905, the evening of the Rector's death,

Account for 'regulating' the organ in 1823

and on the following day work began on installing a new organ made by J W Walker & Sons at a cost of £1,568.

In 1907 an organ case was built in memory of Horace Beck.

Despite being seriously damaged in the Blitz this organ continued in service until the fire which gutted the church in 1975.

When the church was rebuilt it was decided to relocate the new organ in a prominent position in the gallery where it provides a magnificent focal point to the west end of the church.

It was built in 1979 by J W Walker & Sons to specifications drawn up by the then Master of Music, Thomas McLelland-Young.

The case is made from red Canadian pine and was designed by the church architect, Douglas Feast.

An inaugural organ recital was given on 1st June 1979 by the noted organist, Gillian Weir, since when a number of prominent organists have given recitals including Nicolas Kynaston, John Scott and Jane Parker-Smith.

The instrument is a tracker-action organ with 22 stops and three couplers with mechanical action to keys and pedals and electric action for the stops and pistons. It is widely considered to be one of the finest organs in the south of England.

Left: Programme for the inaugural recital on the new organ by Gillian Weir in 1979

Keyboard and pedals of the J W Walker & Sons organ installed in 1979

Small organ used occasionally in the late 1890s and early 1900s

ST. LEONARD'S CHURCH - STREATHAM

THE CHANCEL

The chancel is the space around the altar at the east end of the church. By tradition this area belongs to the Rector, who was responsible for its repair and upkeep. It was here, in past times, where former Rectors would have been buried and the Rector could sell space here to wealthy parishioners who wanted to be buried near the altar, the most sacred place in the church. Hence many past Rectors, Lords of the Manor and other wealthy residents were buried in the chancel of the pre-1831 church.

When St. Leonard's was rebuilt in 1831 a small apse was placed at the eastern end of the building. In 1863 this was removed when the chancel extension was built.

Although the chancel extension is therefore the most recent addition to the church, it was built in an early English style and its design therefore leads people to believe that it is the oldest part of St. Leonard's.

The extension was designed by William Dyce, a noted Victorian artist, who was churchwarden of St. Leonard's from 1862 until his death in 1864 when he was buried in the graveyard. The chancel was later renovated under the direction of the famous architect Benjamin Ferrey FSA who had executed Dyce's original design.

Damage caused to the stonework during the fire of 1975 has added to the chancel's appearance of antiquity and today, unincumbered by a rood screen and choir and clergy stalls, it provides a wonderful vista at the eastern end of the church.

After the fire the stone pillars and arches were cleaned and traces of the original painted designs that once covered them were discovered and can still been seen today. These decorations were designed by William Dyce.

Around the top of the columns of the two central pillars in the chancel can be seen the heads of the four Gospel writers (*) and the four Doctors of the Church (#), who were four eminent Christian teachers who were named in 1298. Some of these heads were very badly damaged in the fire and were reconstructed by Inger Norholt. The northern pillar features on the west: St. Matthew*, north: St. Gregory (Pope Gregory I)#, east: St. Mark* and south: St. Augustine#. The southern pillar features on the west: St. John*, south: St. Ambrose#, east: St. Luke* and south: St. Jerome#.

St. Luke's head on a chancel pillar

The holes drilled into the pillars date from the time of the fire in 1975 when core samples were taken to ascertain if the pillars were still strong enough to support the weight of the building above and that they had not been seriously weakened by the fire.

In the south wall of the chancel is the piscina, a small, shallow basin used for washing the communion vessels. This dates from the time the chancel extension was built in 1863.

The altar is of a simple wooden design and was made by Auker and Sons Ltd., a local carpentry firm based in nearby Mitcham Lane. On Sunday 22nd December 1976 the rebuilt church was dedicated and the new altar was carried into the building by six 'men of St. Leonard's' - George Clay, David McClean, John Carrie, Ian Pratt, Dennis West and Nigel Clifford, the son in law of the church architect, Douglas Feast. The throwover cloth was stitched by Douglas's daughter, Helen, and she, together with other women of the church, placed it in position.

The beautiful altar-frontal, symbolising a phoenix rising from the ashes, was designed and made by Patricia Roberts and was the gift of the architectural practice responsible for the rebuilding of the church, headed by Douglas Feast and his associate, Patricia's husband, David Roberts.

The processional cross and stand was designed by Douglas Feast. The figure of Christ comes from an earlier processional cross which was acquired by the Revd. Michael Hamilton Sharp. When the left arm of Christ was broken off from the cross by accident it was expertly repaired so that now one cannot see the join; the restoration being the gift of Dr Margaret Kesterton.

EAST WINDOW

The east window was unveiled and dedicated on Easter Day 1987. It was designed by John Hayward and features the central figure of Jesus as an amalgam of Christ crucified, risen and reigning.

Christ's deliberately rather awkward pose was designed to suggest a figure at once broken, yet triumphant with his cross becoming his throne.

On either side of Christ are emblems of the writers of the four gospels, the disciples Matthew - an angel (top left), Mark - a lion (top right), Luke - a winged ox (bottom left) and John - an eagle (bottom right).

Below are representations of the sacraments of baptism and communion and above is a dove representing the Holy Spirit.

To obtain minimal obstruction of light through the window John Hayward used a silver stain fired onto the back of the glass to provide the subtle colours featured in this window. This is a technique which dates back to the 14th century, a time when Sir John Ward, whose effigy rests beneath the window, rebuilt St. Leonard's church.

The money for the design and installation of this window was provided by H. M. Customs and Excise in the form of a refund of VAT (Value Added Tax) incorrectly charged on the cost of rebuilding the church. This was obtained as a result of a detailed appeal, successfully lodged with the tax authorities, by the church treasurer, Tom Fortey.

In 1623 reference is made to two coats of arms related to Sir John Ward as being 'in the chancel window, very old'.

When John Aubrey visited the church in 1673 he found stained glass in the east window featuring a red rose with a crown above beneath which was an inscription, then only partially readable: '... Ferechard and Margaret ... wyf glazed ...'. Another window on the south side of the chancel had contained 'some figures, whereof now only that of John the Evangelist remains' beneath which on a gravestone was the inscription 'Here lieth the Body of Mr. Roger Norton, late Subdean of the King's Chapell, and Parson of this Church, which made this Wyndowe of his own Cost, and Deceased the xxvj Day of December in the Yere of our Lord MVCXXVIJ On whos Soule Jhesu have Mercy. Amen'. Roger Norton died on 26th December 1527.

However, in 1814 Manning and Bray refer to the chancel window as containing 'a small whole length of the Virgin; on each side is a red rose, one of which is under a crown' with the aforementioned inscription referring to 'Ferechard and Margaret' and Norton's grave stone being lost.

In 1851 stained glass costing £300 was placed in three windows in the apse which then formed the east end of the church after its rebuilding in 1831. These windows appear to have been discarded when the chancel was extended in 1863, after which a new east window, depicting the *Supper at Emmaus*, was installed in memory of Colonel Alfred Borradaile who died on the 17th May 1861.

This window was destroyed when a bomb fell in the graveyard on 29th October 1940 and was replaced by a new window by Lawrence Lee, installed in 1951, the cost of which was met by the War Damage Commission.

After the fire in 1975 some of the surviving parts of Lawrence Lee's window were included in the new east window (see page 10).

Emblem of St. Matthew

Emblem of St. Mark

Emblem of St. Luke

Emblem of St. John

ST. LEONARD'S CHURCH - STREATHAM

SIR JOHN WARD

Underneath the east window can be found the oldest monument in St. Leonard's church - a mutilated effigy of a recumbent knight in armour dating from c1370. It lies beneath an ornately carved Gothic canopy of a similar date although the two items are placed together for artistic effect and were not originally combined.

At the time John Aubrey visited the church in 1673 the knight was to be found in the belfry where he described it as 'a cumbent figure of white marble in maile with a lyon at his feet which was removed out of the Chancel'.

By the time Manning and Bray published their *History and Antiquities of the County of Surrey* in 1814 it appears to have been relocated within the church, for they advise:

'Upon an alter tomb in the North Wall, under a rich Gothic canopy, lies the mutilated figure of an armed knight, having a pointed helmet, mail gorget, and plaited cuitasses. The canopy is ornamented with quatrefoils; but the pinnacles and some of the other parts are imperfect. It seems probable, from the situation, that it is the tomb of the Founder of the Church, and its form ascertains it to be of the fourteenth century. It has absurdly been called the Tomb of John of Gaunt, who, it is well known, was buried in St. Paul's Cathedral'.

The local legend that the effigy was John of Gaunt was again dispelled in the 1850s when the Revd. Walter Field, a curate at St. Leonard's, discovered a sketch of the monument in an old notebook kept at the Herald's College in London. This was drawn in 1623 before the effigy was damaged.

The sketch shows a recumbent figure in complete armour of the fourteenth century, and described it as placed upon an ancient monument in white marble. The head was covered with a chain-mail cap and rested upon, a pointed helmet; the feet were upon a lion couchant, and the hands clasped as in prayer. With the drawing of the effigy were three coats of arms, two of which are recorded as being in the chancel window and the third 'on a flat marble in the body of the church without any inscription'.

Further research identified the principal coat of arms to be that of Sir John Ward of Surrey. Little is known of Sir John but from the second coat of arms it is believed he married into a Suffolk family.

Subsequent research suggests Sir John may have been a companion of the Black Prince and fought with him at Crecy (1346) and/or Poitiers (1356) in the French Wars. The Black Prince owned land in the parish, being Lord of the Manor of Vauxhall which included Kennington and South Streatham.

It is likely Sir John paid for the rebuilding of St. Leonard's church in the mid-14th century either as a thank-offering for his safe return from the French Wars, or for deliverance from the black death (1348-50), both of which events occurred at around the time of the rebuilding of the church.

Sir John Ward as depicted in the Streatham Window

Interestingly, a list of soldiers who fought in the 100 years war, compiled by the University of Southampton and placed on the internet in 2009, includes a John Ward Esq., a Man at Arms, under the command of John of Gaunt and this may account for the ancient belief that the effigy was connected to the Duke of Lancaster. Ward's captain is given as Lord Edward Despenser who, in the Valor of Edward III, is described as possessing lands in Tooting and Streatham. Hugh le Despenser, who died in 1349, held the overlordship of the Manor of Streatham and Tooting Bec.

Drawing of Sir John Ward's effigy made in 1623

ST. LEONARD'S CHURCH - STREATHAM

Up until the time of the rebuilding of the church in 1831 the knight's armour, including a helmet, breastplate, etc., hung above the monument but these have long since disappeared.

Probably it was also around the time of this rebuilding that a stone coffin was discovered near where the figure was initially placed which could have originally contained the remains of the knight.

After the rebuilding in 1831 Sir John's effigy was relocated under the stairs in the south porch.

In April 1949 it was removed to a position under the Howland monument in the tower, the last remaining part of the old 14th century church which Sir John Ward built. The expense of the removal was met by Ethel Bromhead, widow of the founder-secretary of the Streatham Antiquarian Society, who also paid for the plinth on which the effigy now rests, in memory of her husband, Harold Bromhead.

Harold was a staunch supporter of St. Leonard's church and wrote a number of books on the history of Streatham including *The Heritage of St. Leonard's Church* (1932) and *Streatham's Beginnings* (1936), as well as numerous articles for the parish magazine and the *Streatham News*.

The plinth bears the coat of arms of Sir John Ward and has the following inscription:

'In the 14th century this church was rebuilt by Sir John Ward and this damaged effigy, being a part of the memorial erected to his memory, is placed here in the only remaining portion of the building by Ethel M Bromhead in memory of her husband Harold M Bromhead who passed to his rest 14th April, 1943, in his seventy-fourth year. A devoted servant of this parish'.

During the relocation of the effigy it was discovered it, and the slab on which it rests, were hewn from a single block of alabaster.

Reconstruction of how the effigy originally looked

As to the mutilation of the effigy we know it was intact in 1736 as N Salmon makes no mention of it being damaged in his *Antiquities of Surrey* and yet by the time of Manning and Bray in 1814 it is referred to as being mutilated. Local legend that it was smashed by Cromwellian soldiers during the Civil War therefore has no substance.

I suspect that, while placed in the belfry, it was possibly damaged on one of the occasions the tower was hit by lightning, probably in 1777. Possibly masonry, or heavy timbers, fell on it causing the damage we see today with the lost limbs being smashed into small pieces beyond repair and therefore disposed of. After such damage it was then moved to the north wall of the church, close to the ancient window in which were set Sir John Ward's crests, no doubt where it had originally been located before its transfer to the belfry.

The effigy and canopy were repositioned under the east window by members of the 1st (St. Leonard's) Scout Group in 1977.

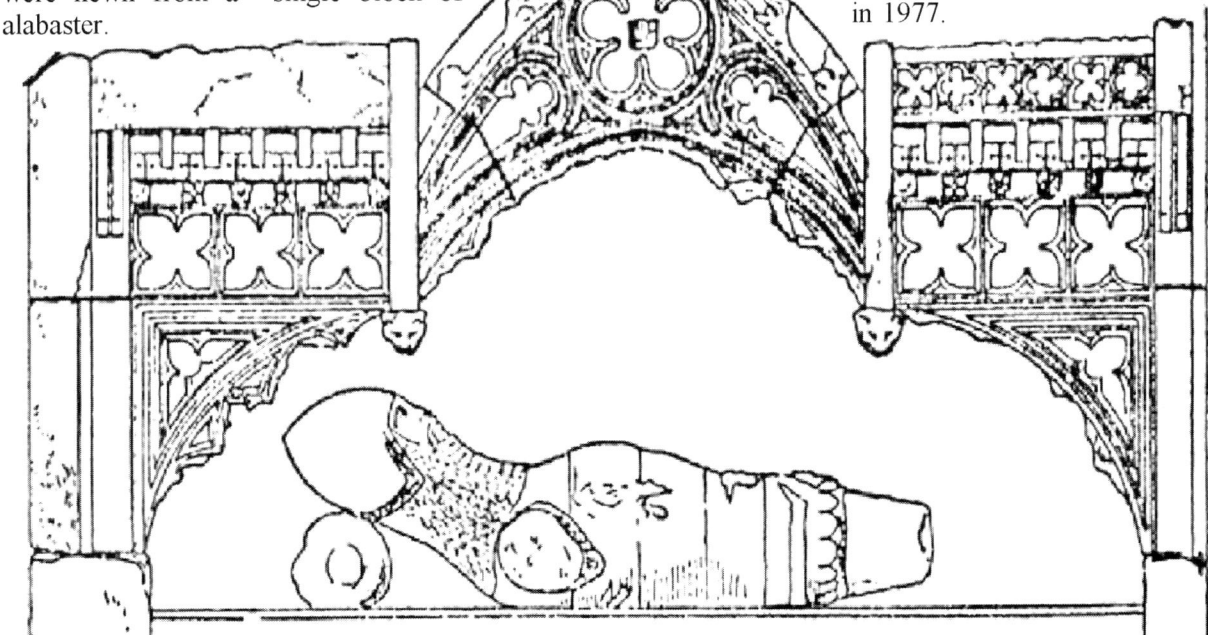

Drawing of the effigy when in the south porch

MOWFURTH BRASS

On the northern wall of the chancel is the fine 16th century memorial brass to William Mowfurth, Rector of Streatham. We do not know when he took up his post here as the registers for 1492-1500 are missing.

However, we do know he was instituted Rector of Puttenham, near Guildford, on 16th October 1506. He resigned this position the following year when he became Rector of Mickleham on 14th February 1507. At the time of his death in 1513 he was holding both livings at Mickleham and Streatham simultaneously.

The brass takes the form of a small full-length effigy, 18¾ inches in height, showing Mowfurth tonsured, with long hair, and wearing the usual mass vestments of the time, very similar to those the clergy of St. Leonard's wear at Holy Communion today.

Below is the inscription in Latin measuring 16½ by 5½ inches, in four lines of text:

'hic iacet dns Willms Mowfurth quondam isti ecclie ac ecclie de Myklam Rector qui obiit xv die Octobris Anno dni millmo vc xiij cuius Aie ppicietur deus amen'. (Here lies William Mowfurth formerly Rector of this church and the church of Mickleham, who died October 15th 1513, upon whose soul may God have mercy. Amen.)

This brass gives us the earliest known image of a Rector of Streatham and would have originally been inset into the stone slab covering his tomb.

The brass appears to have been moved on several occasions over the centuries. The earliest published reference to it appears in John Aubrey's *The Natural History and Antiquities of the County of Surrey* originally prepared in 1673 which records the inscription 'on a brass plate on a gravestone on the south side of the chancel'.

By 1814, when Manning and Bray's *History and Antiquities of the County of Surrey* was published, it was 'fixed in the east wall on the south side of the communion table'.

After the church was rebuilt in 1831 it was placed at the west end of the north aisle and was subsequently relaid in a cement slab, 28 by 22 inches, and fixed to the south wall of the Lady Chapel.

When the church was rebuilt after the fire in 1975 it was placed in its present position where it was a popular attraction for 'brass rubbers' who would place a sheet of paper over it and gently rub the paper with a wax crayon to reproduce the image.

THE NORTH CHAPEL

Before 1975 this part of the church was used as a Clergy Vestry or Sacristy, where the clergy's robes and vestments were kept. The organ was also located here.

On the floor of the chapel are three large tombstones.

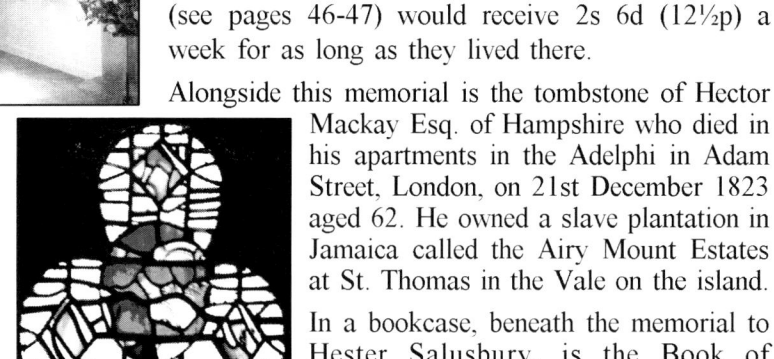

That nearest the east wall is in memory of Major Henniker, who died on 3rd February 1789 aged 33, his wife, Mary, who died aged 47* on 6th February 1803, and their son, Brydges Henniker who passed away on the 17th October 1794 in the ninth year of his age. (*The 4 is poorly carved, leading some sources to record that she died aged 17).

A handsome large, white, marble memorial to Major Henniker used to be on the north wall of the church until it was destroyed in the fire in 1975. This was by the well known sculptor John Bacon, who carved the memorial to Dr. Johnson in St. Paul's Cathedral.

Major was a Christian name, not an army rank. He was the son of the first Baron Henniker, the brother of the second Baron and father of the third Baron.

West of this stone is one of the largest floor memorials in the church in memory of Susanna Arabella Thrale, the second daughter of Henry and Hester Thrale (see page 44). She died on November 5th 1858 aged 88 and was the only one of the Thrale's four surviving daughters who did not marry. She was originally buried in the graveyard but her tomb slab was later incorporated into the floor of the chancel so that all the Thrale memorials would be together. Around the perimeter of the slab can be seen holes which once housed railings that surrounded the grave. The memorial also records the burial of Susanna's nephew, Thomas Arthur Bertie Mostyn, who died at Brighton on 3rd October 1876 aged 75. In 1860 he set up a fund totalling £866 13s 4d (£866.67) so that the 4 almswomen living in the Thrale Almshouses (see pages 46-47) would receive 2s 6d (12½p) a week for as long as they lived there.

Alongside this memorial is the tombstone of Hector Mackay Esq. of Hampshire who died in his apartments in the Adelphi in Adam Street, London, on 21st December 1823 aged 62. He owned a slave plantation in Jamaica called the Airy Mount Estates at St. Thomas in the Vale on the island.

In a bookcase, beneath the memorial to Hester Salusbury, is the Book of Remembrance. This includes the names of all those whose ashes have been interred in the graveyard as well as those who were members of the congregation, or others, who wished to be remembered. The book was dedicated at a special service held on All Souls' Day, 2nd November, 2011.

The book comprises one page for every day of the year and the pages are turned as required so that the names of the departed are shown on the anniversary of their death. The names are also remembered that day in morning prayers and are published in the weekly St. Leonard's church newsletter.

The Book of Remembrance is housed in a specially made wooden case, the design of which replicates the style of the woodwork panelling front to the gallery. The case was presented to the church in memory of Gobind Parkash Sehgal.

This book replaces an earlier Book of Remembrance instituted after the fire in 1975.

Almost hidden from view, high above the Hoare memorial (see page 46), and beneath the ridge of the roof, is a small stained glass window which was the gift of the firm of Sawyer and Fisher of Epsom, the Quantity Surveyors involved in the rebuilding of the church in 1975. This window was made from salvaged pieces of broken stained glass recovered from the debris after the fire.

Quantity Surveyor's window

Book of Remembrance

Hector Mackay's Tombstone

THE STREATHAM WINDOW

After the chancel was extended in 1863 a Sacristy was created in this part of the church to accommodate the robes and vestments of the clergy.

In 1865 James Brand, the father-in-law of William Dyce (see pages 42 and 61), erected two stained glass panels here representing Charity and Humility.

On the northern wall of the old Sacristy, the two windows now above the candle stand, had stained glass representing the Epiphany placed in them in 1875 in memory of Eliza and Adelaide Haigh.

All these windows were lost when a bomb exploded in the graveyard on the 29th October 1940 and for the next 20 years they were filled with plain glass.

In 1960 J. W. C. Potts and family erected stained glass in the east windows in memory of their parents. This depicted the story of the Good Samaritan and was designed by Frederick Cole. This window was destroyed in the 1975 fire.

In 1980 a new east window was dedicated which is now known as the Streatham window. It was designed by John Hayward and was the bequest of Connie Measures as a memorial to her and her husband, Ron Measures.

Through its many images it tells part of the fascinating history of the church and parish from Roman times to the present day.

At the very top of the window is a roundel the central feature of which is a crossed Bishop's crook and a cross behind which are two linked manacles encompassing the letters S, on the left, and L on the right. The letters stand for St. Leonard, the patron saint of prisoners, and hence the manacle chains in the motif. The top left segment of this window has the arms of the diocese of Winchester, of which Streatham was part until 1877; top right are the arms of the diocese of Rochester which included Streatham between 1877 and 1905; and at the bottom are the arms of the diocese of Southwark, of which St. Leonard's has been a member since 1905.

At the top of the left hand window are representations of the Roman coins found in St. Leonard's churchyard bearing the heads of the Emperors Carausius (died 293AD), Constantius Chlorus (died 306AD) and Constantine the younger (died 340AD).

These coins, together with other Roman relics found in the area, including a votive figure lof the Goddess Venus discovered when digging the foundations for the church of the English Martyrs in Tooting Bec Gardens in 1892, and remains of a Roman pavement discovered when digging the

ST. LEONARD'S CHURCH - STREATHAM

foundations for the Bedford Park Hotel in 1882, indicate a Roman settlement at Streatham from at least the late third century.

Working down the window we next come to the seated figure of the Abbot of Chertsey on the left. The earliest mention of 'Totinge cum Streatham' appears in a grant of lands to Chertsey Abbey in 675 which was confirmed several times over the following centuries up to, and including, the year 953. Beneath the Abbot's right foot is a representation of the seal of the Abbey.

Sitting next to the Abbot is St. Anselm, Abbot of Bec, and later Archbishop of Canterbury. He holds on his lap a representation of the first St. Leonard's church. The earliest reference to a church at Streatham is recorded in the Domesday survey of 1086 which lists 'A chapel which pays 8s' (40p). The Abbey of Bec named this chapel after the patron saint of their Abbot, Saint Leonard, who was widely honoured in Normandy. Behind the two abbots are the three red chevrons featured on the arms of Richard fitz Gilbert, Lord of Clare, who endowed the manors of Streatham and Tooting Bec on the Abbey.

To the right Sir John Ward is seated holding his sword in his right hand and, in his left, a banner on which is his coat of arms. On his lap is a representation of St. Leonard's church which he rebuilt c1350. All that remains today of Sir John's church is the lower part of the tower, its arch and the spiral staircase leading up to the belfry. Behind Sir John are the arms of his friend, the Black Prince, with whom he probably fought at the battles of Crecy and Poitiers. (see page 36).

To the right, below Sir John's standard, are 3 lilies which form part of the crest of Eton College, which held the Manor of Streatham and Tooting Bec as a gift from Henry VI in 1442 as part of his endowment, until reclaimed by Edward IV.

The figures to the left, spanning both windows, represent the Brabourne incident. In 1394 John Brabourne, fleeing his master, sought sanctuary in St. Leonard's church. His three pursuers violated the sanctuary dragging him from the church to Guildford Prison. The Bishop of Winchester, William of Wykeham, appealed to Richard II, who ordered the three men to bring Brabourne back. Then for three consecutive Sundays, stripped to shirts and drawers and carrying lighted tapers, they had to walk through the village of Streatham while the Rector, John Elslefeld, whipped them with a rod and announced their guilt, after which they had to kneel in the nave at high mass, repeat the Magnificat and pray for forgiveness.

To the left of the penitents is Edmund Tylney, Master of the Revels to Queen Elizabeth I and King James I, and censor of William Shakespeare's plays. He was buried in St. Leonard's church in 1610 and his monument is in the Chapel of Unity (see page 20).

Beneath Edmund are Hester and Henry Thrale against the background of Streatham Park, their home which was the centre of cultural and social life in late 18th century Streatham. Their family crest is below and to the left is Dr. Johnson, the famous writer who, between 1766 to 1782, frequently stayed at Streatham Park. To the right is James Boswell another visitor to the Thrale's home here and biographer of Dr. Johnson (see page 44).

ST. LEONARD'S CHURCH - STREATHAM

To the right is John Howland, above whom are his family's coat of arms. John was Lord of the Manor of Streatham and Tooting Bec between 1679-1686 and was buried in St. Leonard's church. His handsome memorial is on the north wall of the tower (see page 12).

Below is the coat of arms of the Dukes of Bedford who became Lords of the Manor of Streatham and Tooting Bec in 1695 through the marriage of the 14-year-old Wriothesley Russell, Marquess of Tavistock, to the 13-year-old Elizabeth Howland, the daughter of the aforementioned John Howland.

To the left of the Bedford arms, spanning both windows, is a horse and carriage reminding us of how fashionable Streatham was in the late 18th and early 19th centuries when St. Leonard's attracted visitors who would travel many miles to worship here.

To the left of the Thrale family crest is an image of the church at this time showing St. Leonard's before it was rebuilt in 1831.

Beneath the aforesaid images is a roundel containing illustrations of seven of the 16 daughter churches built within the old parish of Streatham. Commencing on the left, going anti-clockwise, those featured with the dates they were built in brackets are St. Margaret the Queen, Streatham Hill (1900); St. James, Mitcham Lane (built in two parts 1912-1916); Holy Redeemer, Streatham Vale (1932); St. Andrews, South Streatham (1886); Christchurch, Streatham Hill (1841); behind is St. Peters, Leigham Court Road (1870) and Immanuel Church, Streatham Common (1854 and enlarged and rebuilt as shown 1865).

Beneath the churches on the left is a small dedication reading 'In memory of Ronald and Constance Measures 1985' and on the right an illustration of the Dyce fountain to be found on Streatham Green and the initials WD standing for its designer William Dyce. This famous Victorian artist is credited with designing the Chancel extension in 1863, the work being executed by the architect Benjamin Ferrey. Dyce was churchwarden of St. Leonard's from 1862 until his death in 1864 and is buried in the churchyard (see page 61).

Opposite we see St. Leonard's church engulfed in flames on the night of Monday 5th May 1975, when the church was gutted leaving only the tower and outer walls of the building standing the following morning.

Above the flames on the right is a bishop's mitre behind which are crossed crooks and the initial M with the dates 1959-1980. This represents Mervyn Stockwood, the Bishop of Southwark between these dates, who was a parishioner of this church as the bishop's residence is situated in Tooting Bec Gardens.

In the left hand corner of this panel is a roundel containing plans for the new church together with architect's instruments and the initials DF standing for Douglas Feast, of 11A Streatham Common South, whose partnership designed the church we see today.

Douglas Feast

In the right hand corner we can see the then Rector of Streatham, Michael Hamilton Sharp, kneeling in prayer, with his initials MHS behind. He was Rector of Streatham between 1972-1981 encompassing the period of the fire and rebuilding of the church. He died on 29th January 2010.

Lastly, at the very bottom of the window is John Hayward's signature and the date 1980, when he designed and made the window. Currently all the stained glass in St. Leonard's church is by him, although each is of a very different style.

John Hayward was born on 17th July 1929 in Tooting and attended Bec Grammar School which subsequently became part of Ernest Bevin school. During his lifetime he made almost 200 windows which illuminate churches and cathedrals throughout Britain and abroad. His first window was designed and made in 1955 for Christchurch, Streatham Hill, in which parish he was an active member at that time. It was given on the occasion of his daughter's baptism and his own confirmation and shows St. Michael crushing down the devil under his feet. John was to make two more windows for Christchurch in 1961 and 1981. His first major commission was a scheme of windows for the ruined Wren church of St. Mary-le-Bow in London. Other examples of his work can be found in Blackburn and Norwich Cathedrals and at Sherborne Abbey where he designed a replacement for the Victorian Great West Window, which was dedicated in the presence of the Queen in 1998. John died on the 19th May 2007 aged 77.

William Dyce

ST. LEONARD'S CHURCH - STREATHAM

ICON OF ST. LEONARD AND ST. LAURA

In the north chapel, under the Streatham window, is an icon of St. Leonard with St. Laura and a prisoner. It was presented to the church by the congregation in memory of Laura Margaret Wilcox (1974-1995), the eldest daughter of the then Rector of Streatham, Revd. Jeffry and Mrs. Claire Wilcox. Laura died on 30th October 1995 from a brain tumour at the tragically early age of 21.

The icon was blessed at a special service on All Saints' Day (1st November) 1998.

It was painted by Leon Liddament (1943-2010), a member of the Russian Orthodox Brotherhood of St. Seraphim at Little Walsingham, in Norfolk.

St. Leonard is depicted in the central panel in the Byzantine style wearing Deacon's robes with an Abbot's staff and his fingers poised in blessing. In his left hand he holds the Gospels and an incense burner, the chain of which carries a broken lock, a reminder that he is the patron saint of prisoners.

St. Leonard is thought to have been born in Orleans, France. He was the son of an army officer and courtier of King Clovis I of France who ruled from 481-511. At Leonard's baptism, on Christmas Day 496, the King stood as his godfather and the service was conducted by St. Remy, Bishop of Rheims.

Despite his privileged upbringing, neither the court, nor the army, held any attraction to him and on reaching manhood Leonard entered the monastery at Miscy, afterwards St. Mesmin, near Orleans. When Leonard was 40 he left Miscy and went to Noblac, located in the dense forest near Limoges, where he lived as a hermit.

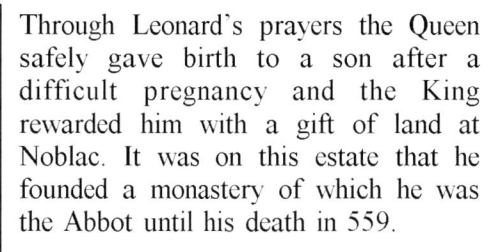

St. Leonard

Through Leonard's prayers the Queen safely gave birth to a son after a difficult pregnancy and the King rewarded him with a gift of land at Noblac. It was on this estate that he founded a monastery of which he was the Abbot until his death in 559.

Leonard was particularly concerned about the plight of prisoners and captives and persuaded the King to grant him the privilege of freeing those he considered worthy of release when he visited local jails. It is for these acts of mercy he is mainly remembered today and he subsequently became the Patron Saint of prisoners and captives and women in labour.

Leonard died on November 6th 559 with the words 'Lord Jesus receive my soul' on his lips. He was buried in the little church he had built at Noblac.

St. Leonard was the patron saint of the Abbot of Bec Abbey in France and our church was dedicated to him after the manors of Streatham and Tooting Bec were conferred on the Abbey by Richard fitz Gilbert, Lord of Clare, who received them from William the Conqueror after the Battle of Hastings in 1066.

Today there are 177 churches in England dedicated to St. Leonard.

To the left of St. Leonard in the icon is a prisoner wearing shackles, in a gesture of intercession.

To the right is St. Laura, depicted holding the martyr's cross. She was born in Cordoba, Spain. After her husband died she became a nun at Cuteclara and later an Abbess. She was captured by the Moorish conquerors who scalded her to death by placing her into a cauldron of molten lead in 864.

St. Laura was the name saint of Laura Wilcox after whom the icon is a memorial.

HENRY THRALE MONUMENT

The first of the Thrale family monuments on the south wall of the church is that in memory of Henry Thrale, the famous Southwark brewer and MP, who died on the 4th April 1781 aged 52.

Henry Thrale was born c.1728 at the Alehouse in Harrow Corner, Southwark, and was the son of a wealthy brewer, Ralph Thrale and his wife Mary. He was educated at Eton and Oxford.

On the death of his father in 1758, he inherited the family brewery and built up the business to become the fourth largest brewery in London.

He also inherited the family's country mansion at Streatham Park which his father, Ralph, purchased in 1735 and which Henry subsequently extended and enlarged.

On the 11th October 1763 he married Hester Lynch Salusbury, at St. Anne's Church, Soho; he was in his mid 30s and his bride 22. This was an arranged marriage which Hester agreed to 'not on passion but on reason' without ever having been alone in Henry's company. The couple had twelve children, although only four daughters survived into adulthood.

In 1765 Henry was elected MP for Southwark, the constituency in which his large brewery was located.

Streatham Park, or Streatham Place as it was then known, was the fashionable country residence of the Thrales. Here, Henry Thrale together with his wife, Hester, entertained the leading luminaries of the day, including such famous men as David Garrick, Edmund Burke, Oliver Goldsmith, Sir Joshua Reynolds, and Dr. Samuel Johnson, the great 18th century lexicographer and author.

Johnson was a frequent visitor to Streatham Park between 1766 and 1782 and would often spend the middle part of the week there as the Thrale's guest. Hester ensured a room was always available for him and he enjoyed the full facilities of the house, including its large library. In the summer months he would pass many a pleasant day strolling round Streatham village or quietly sitting in the summerhouse which stood in the extensive gardens that surrounded the house. (The Summerhouse was later removed to Kenwood House in Hampstead where it was destroyed by vandals in 1991).

Henry Thrale died on 4th April 1781 after suffering a series of apoplectic fits after eating an enormous eight course meal washed down 'with strong beer in such quantities the very servants were frighted'. He was buried in his family vault in St. Leonard's Church on 11th April.

Henry Thrale

Following Thrale's death, Dr. Johnson assisted at the sale of the brewery and when asked what he thought the value of the business was replied 'We are not here to sell a parcel of boilers and vats, but the potentiality of growing rich beyond the dreams of avarice.'

Mr. David Barclay purchased the business for £135,000 and in partnership with the brewery's manager, John Perkins, the brewery subsequently became known as Barclay & Perkins. In 1955 it amalgamated with Courage and Co.

As a result of Henry's overeating in the final years of his life he put on considerable weight and his huge coffin can be seen in the crypt of the church where its gigantic size is evidence of Henry's great girth caused by his love of good food and drink (see page 55).

After Henry's death relations between Hester Thrale and Dr. Johnson cooled, especially following Hester's subsequent marriage to her family's Italian music teacher, Gabriel Piozzi.

In 1782, Dr. Johnson made his final visit to his 'second home' at Streatham Park, and with a heavy heart wrote in his diary later that evening 'Sunday, went to church at Streatham. I bade farewell to the church with a kiss.'

Hester Thrale

ST. LEONARD'S CHURCH - STREATHAM

Hester Piozzi and her new husband only lived at Streatham Park briefly and the house was subsequently let to a number of wealthy tenants, including the Prime Minister, Lord Shelburne, who rented it between 1782-5. By the early 1860s the house had became neglected and was in a poor state of repair. The estate was eventually sold and in 1863 the house was demolished and the surrounding land sold off for residential development.

Henry Thrale's monument was cut by the famous British sculptor, John Flaxman, for which he charged £31 13s 4d (£31.66p). (Flaxman also carved the adjoining memorial to Sophia Hoare (see page 46).

The Latin inscription was composed by Dr. Samuel Johnson. He wrote only four epitaphs - one intended for Hogarth's tomb which was never used; and one for the monument in Westminster Abbey for Oliver Goldsmith, the great novelist, poet and playwright. The other two are here in St. Leonard's church being the inscription for Henry Thrale's mother-in-law, Hester Salusbury (see page 48) and the one which adorns Henry Thrale's memorial.

Johnson's epitaph for Henry is fulsome in its praise of his good friend but is so great in length that for it all to be included on the tablet it has been carved in the smallest of letters.

An English translation of the Latin text reads:

'Here are deposited the remains of HENRY THRALE who managed all his concerns in the present world, public and private, in such a manner as to leave many wishing that he had continued longer in it, and all that related to a future world as if he had been sensible how short a time he was to continue in this. Simple, open and uniform in his manners, his conduct was without either art or affection. In the Senate attentive to the true interest of his King and country, he looked down with contempt on the clamour of the multitude; although engaged in a very expensive business he found some time to apply to polite literature; and was ever ready to assist his friends labouring under difficulties with his advice, his influence, and his purse. To his friends, acquaintances and guests he behaved with such sweetness of manners as to attach them all to his person; so happy in his conversation with them as to please all, though flattered none.

Born 1728 - Died 1781

In the same tomb lie interred his father, RALPH THRALE, a man of vigour and activity, and his only son HENRY, who died before his father, aged 10 years. Thus a happy and opulent family raised by grandfather and augmented by the father became extinguished with the grandson. Go, Reader, and reflect on the vicissitudes of all human affairs, meditate on Eternity.'

This monument was smashed into 32 pieces in the fire and was restored by Inger Norholt, the money for the restoration being donated by the Johnson Society.

When St. Leonard's was rebuilt in 1831 the coffins in the Thrale family vault were transferred to chamber 7 in the crypt. Hester Thrale was buried alongside her second husband, Gabriel Piozzi, at Tremeirchoin Church in Wales.

Dr. Samuel Johnson

Streatham Park, the home of Henry and Hester Thrale

SOPHIA HOARE MONUMENT

The central of the three Thrale family monuments on the north wall of the church is that in memory of Sophia Hoare, the seventh child of Henry and Hester Thrale.

She was born on the 23rd July 1771 and christened in St. Leonard's church a couple of weeks later on August 11th.

Sophia married when she was 36 taking as her husband the prosperous banker, Henry Merrik Hoare. They were wed at St. Marylebone's church in London on the 13th August 1807.

Henry was a partner in the family bank, C Hoare & Co., and was a grandson of Sir Richard Hoare, who was Lord Mayor of London in 1745 and himself was a grandson of the bank's founder, Sir Richard Hoare who was Lord Mayor of London in 1712.

It was a great sadness to the Hoares that they never had any children, despite which their marriage was a happy one spent at Henry's home at 31 York Place in London (now 109 Baker Street).

Of the Thrale's twelve children only four daughters survived into adulthood of which Sophia was the first to die on 8th November 1824 aged 53.

The great British sculptor, John Flaxman (1755-1826) was engaged to carve Sophia's monument. This features her husband on the left in deep grief at his wife's death, with her three surviving sisters, Hester Maria (Queenie), Susanna Arabella and Cecilia Margaretta, mourning at her feet and an angel pointing heavenward above her draped body. This was one of the last monuments carved by Flaxman and cost £200. He completed it when he was 70 and died the following year. Other examples of his work are to be found in Westminster Abbey and in St. Paul's Cathedral where he made the monument in memory of Admiral Lord Nelson.

In 1832, Henry Hoare and Sophia's surviving three sisters, paid for the building of the Thrale Almshouses which provided subsidised accommodation for four poor widows or single women who had 'attained an honest old age' in Streatham. They were built in memory of the sisters' parents, Henry and Hester Thrale of Streatham Park and stood on a site now occupied by the Pratts and Payne public house at 103-105 Streatham High Road.

In 1930 the almshouse site was sold for £11,792 and the money was used to build eight new almshouses in Polworth Road which were designed by Cecil M

Sophia Hoare

ST. LEONARD'S CHURCH - STREATHAM

Thrale Almshouses, Streatham High Road, c1910

Thrale Almshouses in Polworth Road in 2010

Quilter. Following the death in 1938 of Lady Edith Robinson, the wife of the president of the Streatham Conservative Association, it was decided to erect an additional almshouse in her memory at the Polworth Road site. A public collection raised £1,000 which covered the cost of two additional almshouses that were opened in 1940.

In 2014-15 the Thrale Almshouses were renovated and extended. The block erected in memory of Edith Robinson was demolished and this site, together with vacant land facing Polworth Road, was redeveloped with two additional blocks of almshouses. The site now provides updated, modern, accommodation for up to 17 women over the age of 60 in 15 two-bed, and 2 one-bed, flats.

Sophia's memorial was badly damaged in the fire in 1975 but was subsequently restored by the Danish restorer, Inger Norholt; the money for the restoration being donated by the Johnson Society. It is one of the artistic treasures of St. Leonard's being one of the few examples of Flaxman's work to be found in a London suburban church.

The inscription on the memorial reads:

'Blessed are the Dead who die in the Lord,
Yea, saith the Spirit,
that they may rest from their Labours
and their works do follow them'
Rev c14 v13
Sacred to the memory of
SOPHIA
wife of HENRY MERRIK HOARE ESQ
of London
third daughter of HENRY THRALE ESQ
and granddaughter of
HESTER MARIA SALUSBURY
whom in her recorded virtues she equalled
and the excellence of whose mind
was expressed in the beauty of her countenance
Born 23rd July 1771
Died 8th November 1824

Underneath the text is a small replica of the Hoare family crest.

The inscription refers to Sophia as the 'third daughter of Henry Thrale' which is not strictly correct as three of her elder sisters died in infancy and this wording refers to the surviving four daughters.

It is interesting to note no mention is made in the inscription to Sophia's mother, Hester Thrale, with whom she had a strained relationship following Hester's marriage to her daughter's singing teacher, Gabriel Piozzi, and their subsequent adoption of her husband's nephew, John Salusbury Piozzi.

The Thrale family crest and the plaque commemorating the erection of the original Thrale Almshouses in 1832

HESTER MARIA SALUSBURY MONUMENT

Hester Maria Salusbury was the mother of Henry Thrale's wife, Hester Thrale. She was born in 1707, the daughter of Philadelphia Lynch, the former Lady Cotton.

In 1739 she married her cousin, John Salusbury, at St. Paul's Cathedral. The Salusbury's had only one child, their daughter Hester Lynch Salusbury, who was born in 1741.

John experienced periods of financial difficulty as he not only had to support his own family, but also his brother Tom, later to become Sir Thomas Salusbury (a judge in the Admiralty Court), and his mother and his younger brother, Harry who, after an accident, became 'an idiot'.

John made two unsuccessful visits to Nova Scotia in Canada in a bid to increase his family fortunes and his periodic financial problems probably accounted for his irascible and quick tempered nature.

It was Sir Thomas who proposed Henry Thrale as a suitable match for his niece. He had met Thrale in London and had hunted with him, probably in Croydon where Thrale kept a pack of fox hounds and had a hunting box. After meeting Henry, Hester Maria was in complete agreement with the match but when her husband, John, learnt of the plans he was furious and in one of his customary rages refused to sell his daughter 'for a barrel of porter' to a brewer!

This impasse was resolved when John suddenly died in December 1762 and the couple were wed at St. Anne's Church in Soho on 11th October 1763.

At Streatham Park Hester Maria had numerous clashes with Dr. Johnson and for many years neither party liked each other. No doubt at the centre of this animosity was their wish to dominate the affections of the young Hester Thrale.

However, as the years passed, and especially after Hester Maria developed breast cancer, an illness which she bore with great fortitude and courage, Johnson came to respect her, and she him as she appreciated how he would support her daughter after her death. This respect grew into a genuine affection for each other.

Hester Maria died, aged 66, on 18th June 1773 and was buried in the Thrale family vault at St. Leonard's church on 24th June.

As a token of the love Dr. Johnson had for Hester Thrale, and the growing good will he had come to feel towards Mrs. Salusbury, he agreed to write her epitaph.

The engraver, Joseph Wilton RA, complained that Johnson's proposed inscription of 570 letters was far too long and should be cut back to 350 letters, or twelve lines of text. However, Hester Thrale was not prepared to shorten the epitaph and Wilton agreed to solve the problem by carving smaller letters over 17 lines.

An English translation of the Latin text reads:

Hard by is buried Hester Maria, Daughter of Thomas Cotton of Combermere, Baronet, of the County of Cheshire, wife of John Salusbury, Gentleman of the County of Flint. In person charming, charming too in mind, agreeable to all at large, to her own circle very, very loving, so highly cultivated in language and the fine arts that her talk never lacked brilliancy of expression, ornateness of sentiment, sound wisdom and graceful wit. So skilled at holding the happy mean that amid household cares she found diversion in literature, and among the delights of literature diligently attended to her house affairs. Though many prayed for length of days for her, she wasted away under a dread cancer-poison, and as the bonds of life were greatly loosed, passed away from this earth in full hope of a better land.

Born 1707. Married 1739. Died 1773.

Hester Maria Salusbury

This monument was restored after the fire by Inger Norholt, the money for the restoration being donated by the Johnson Society.

Johnson's epitaph drew considerable praise. In 1784 it was discovered that it had been copied verbatim, other than for the name and dates, and used on a memorial to a lady in Walthamstow for which a local man had been paid for its composition. When this plagiarism was made known to the gentry of the parish they 'fairly hissed the imposter out of their town' and the monument was removed from the church.

THE LADY CHAPEL

Above the altar is a small, carved crucifix. This was acquired by the Revd. Jeffry Wilcox from the L'Arche Community based at 9-13 Norwood High Street and was made by one of their members. L'Arche was founded in 1973 and currently operates ten Communities in the UK providing homes to more than 100 people with learning disabilities.

Resting on a covered stand, to the right of the altar, is an icon showing the Virgin Mary and the baby Jesus. This was purchased by the Revd. Jeffry Wilcox from Christ Church in North Brixton and was painted by a local artist living in that parish.

At the foot of the stand on which the icon is placed are the remains of small marble arch with an angel with its hands clasped in prayer on the front. This is one of a pair, the other being in the Chapel of Unity, which Revd. Michael Hamilton Sharp obtained from St. James's Church, Piccadilly, as an adornment for St. Leonard's.

On the floor of the chapel, to the north west of the altar, is a large memorial stone to Richard and Mary Wilson who lived in the parish of St. Andrew's Undershaft in the City of London. Mary died on the 5th January 1741 aged 59 and her husband, Richard, passed away some three months later on the 14th April 1741 aged 63. The stone also commemorates their daughter, Ellin Demee, wife of Daniel Demee. She died aged 31 on 18th March 1750.

To the west of this stone is a memorial to Sarah, wife of John Thurlin, who died aged 26 on 2nd December 1755. Despite her young age the stone tells us that she had 9 children and was a 'good wife and sincere friend' to her husband. The stone also commemorates her daughter, Susanna, who died aged 5 on 18th August 1755 and her son, William, aged 4 who died on 28th October 1755. The family lived in Oxford Road in the parish of St. George's, Hanover Square, in London where John ran a corn chandler's business. 1755 must have been a tragic year for the family for not only did John lose his wife, but also two of their children, all within the space of four months. The inscription on this stone ends with a poignant and familiar verse:

Mourn not for me
　my husband dear
I am not dead
　but sleeping here
As I am so must you be
Therefore prepare
　to follow me

In 1757 John married again, taking as his second wife Mary White of St. Marylebone. In 1762, his corn chandler's shop in Oxford Road was involved in a tragic accident when two men called Harper and Anderson were throwing a 7lb weight for a bet in the street outside. Harper, in attempting to throw the weight farther than Anderson, misjudged his throw and the weight hit the door post of John's shop and rebounded off the building to hit a woman on her head who was standing nearby who subsequently died from the blow.

The Lady Chapel in the 1920s

ST. LEONARD'S CHURCH - STREATHAM

LADY CHAPEL WINDOW

This window was designed by John Hayward and was dedicated in 1983. It was partly funded by the sale of a parish cookery book containing recipes from church members, both male and female.

Each section of the window tells part of the gospel story from the Annunciation to Pentecost.

Top left we see Mary and John the Evangelist standing by the Crucified Christ on Calvary on Good Friday. Above, in green, is a representation of the curtain in the temple in Jerusalem that was torn in two at the moment of Christ's death.

Top right features Pentecost with the Virgin Mary and the twelve disciples of Jesus receiving the Holy Spirit symbolised by the white dove descending from above them. Pentecost is celebrated seven weeks after Easter Sunday, the time of Christ's resurrection.

The two central panels depict Christ's birth and feature well-known elements of the Christmas story. On the left the three Magi bring their gifts of gold, frankincense and myrrh while above them shines the star which led them to Bethlehem. In the centre are the shepherds with their sheep visiting the stable where Christ was born, featured on the right, wherein Mary and Joseph are depicted with the baby Jesus.

Bottom left features the Annunciation with the angel Gabriel telling the Virgin Mary that she would conceive and become the mother of Jesus, the Son of God. Above them a white dove symbolises the descent of the Holy Spirit and behind is part of the text from Luke, chapter 1 verse 28 *Hail, thou that art highly favoured, the Lord is with thee: blessed art thou among women.* The Feast of the Annunciation is celebrated on March 25th.

Bottom right shows Mary, on the left, with her cousin Elizabeth at their meeting shortly after the Annunciation, at which time Elizabeth was 6 months pregnant, and whose baby was to become John the Baptist. Above them is a paschal lamb, a visual representation of Jesus as the Lamb of God, dating from the Middle Ages.

Behind is part of the text from Luke, chapter 1 verses 46 and 47 *My soul doth magnify the Lord and my spirit hath rejoiced in God my Saviour.* These are the opening verses of the Magnificat (Luke c1v46-55) which Mary speaks to her cousin, Elizabeth, when they meet.

Formerly these windows contained pictures representing Faith and Hope erected here, shortly after the building of the chancel extension in 1863, in memory of William Haigh of Furzedown House who died 29th March 1861.

These windows were destroyed by the bomb that fell in the graveyard on 29th October 1940 and were replaced in 1951 by windows designed by Lawrence Lee, paid for by the War Damage Commission.

Lee's windows showed the Virgin Mary and the baby Jesus receiving the gifts of the Magi. From above, the divine grace descended upon Mary.

These newer windows were destroyed in the fire which devastated the church in 1975.

ST. LEONARD'S CHURCH - STREATHAM

THE BLACK MADONNA

The 'Black Madonna' is one of the poignant reminders of the fire in 1975. The day after the fire the black, charred remains of the carving of the Virgin Mary which used to stand on the top of the rood screen was rescued from the ashes by David and Betty Ashdown. Although this was badly scorched by the flames the carving was recognisable. It was renamed the 'Black Madonna' and now stands in the Lady Chapel of the rebuilt church.

The carved oak rood screen, separating the chancel from the nave, was presented to St. Leonard's church in June 1915 by Arthur James Parker in memory of his father, Henry, who died on 4th July 1887.

The screen was dedicated by Bishop Hook, a former Bishop Suffragan of Kingston, at a special service attended by the Mayor and Corporation of Wandsworth, in which borough Streatham was then located.

The rood screen was designed by Harold C. King of Westminster in the style of the 15th century screens which are often to be found in East Anglican churches. It comprised eight bays, the central two bays forming an entrance into the chancel.

The Rood Screen in the 1930s showing the crucifix with the Virgin Mary and John the Evangelist

In 1925 Arthur donated the rood figures in memory of his mother, Charlotte. These were mounted on top of the screen and comprised a central crucifix flanked by the Virgin Mary and St. John the Evangelist.

The following year he gave the parclose screen between the choir and Lady Chapel in memory of his brother, Walter James Parker.

Through all these memorials to members of his family Arthur dramatically changed the view of the chancel from the nave and provided St. Leonard's with some of its largest memorials.

The Parker family moved to Streatham in the mid 1880s and lived in a large Victorian house called Fairlight in Oakdale Road.

Following Henry's death, the family moved to Gleneagle House, No. 1 Gleneagle Road.

Arthur was a staunch supporter of St. Leonard's church and was vice-president of the Church Council, chairman of the Parochial Finance Committee and a trustee of Streatham Hall.

He was chairman of Parker, Wellesley and Company Ltd., dried fruit merchants of Lovat Lane, in the City of London.

Arthur never married and died at his home at 102 Gleneldon Road in January 1945 aged 80 and was buried in the family grave at Honor Oak cemetery.

ST. LEONARD'S CHURCH - STREATHAM

> HERE LYETH THE BODY OF ANNE LIVESEY ELDEST
> DAVGHTER TO THOMAS CROMPTON OF BENNINGTON
> IN THE COVNTY OF HERTFORD ESQVIER WIFE TO
> GABRIELL LIVESEY WHO DYED IN CHILDBED THE
> LAST DAY OF IANVARY 1598. WHOSE SOVLE IS AT
> REST WITH GOD   ÆTATIS SVÆ 20.

LIVESEY BRASS

On the south wall of the Lady Chapel is the latest of the three brasses to be found in St. Leonard's church. This dates from 1598 and is in memory of Ann Livesey who died on 31st January.

In 1981 this brass was discovered in the crypt by youngsters working on a Lambeth Community Council restoration scheme. Somehow, after the fire in 1975, no one had recognised it as one of the ancient memorials of St. Leonard's. Originally it was situated on the west wall of the south aisle of the church and was placed in its present position in 1981.

The inscription reads :

Here lyeth the body of Ann Livesey eldest daughter to Thomas Crompton of Bennington in the County of Hertford Esquire wife to Gabriell Livesey who died in childbed the last day of January 1598 whose soule is at rest with God Aetatis svae 20

As the inscription tells us Ann died at the tragically young age of 20 'in childbed'. As the registers do not detail any baby buried at this time one must assume her death was caused by complications of her pregnancy.

Gabriel Livesey

Eight months earlier she had married Gabriel Livesey, the third son of Robert and Elizabeth Livesey of Tooting Bec. Elizabeth was the widow of Robert Pakenham, Lord of the Manor of Streatham and Tooting Bec. Following Elizabeth's death on 29th October 1573 Gabriel's father married Amy Hobbes, the widow of William Hobbes.

Robert Livesey played a prominent role in local society and was a Justice of the Peace and was twice high sheriff of the County of Surrey. He died on 21st August 1608, aged 81, and was described on his monument in St. Leonard's church, which sadly was destroyed in the fire, as a man of 'approved wisedome, integritie, courage, and industrie'.

Amy Livesey died on 21st November 1617, aged 76, and the monument recorded that she was 'a memorable matron for pious devotion, charitie, hospitality, &c.'

Their monumental inscription ended with the words:

The Memorial of the Just shall be blessed,
LIVESAYE the name, God here them gave,
And now Lives-aye indeed they have.

Amy was a generous benefactor leaving in her will £3 per annum to be distributed among the poor of the parish at Easter and Christmas. In addition she bequeathed the sum of 13s 4d (68p) to be paid in perpetuity for the preaching of a sermon on Easter Day in St. Leonard's church.

Gabriel Livesey was baptised in St. Leonard's church on 25th July 1566 and was probably born in the family home at

The monument to Gabriel Livesey, and his second wife, Ann, in Eastchurch Parish Church, Kent

Revd. Michael Hamilton Sharp with the rediscovered Livesey brass in 1981

ST. LEONARD'S CHURCH - STREATHAM

Broadwaters, a large house the site of which is now occupied by Knapdale at 21 Tooting Bec Road.

Following the death of his first wife, Ann, in 1598, Gabriel subsequently married Ann, the daughter of Sir Michael Sondes of Throwley in Kent. He and his second wife lived at Eastchurch, Kent and Gabriel was a High Sheriff of the county in 1618. There is a magnificent monument to them in the local parish church on which both their effigies lie full length on their tomb.

In 1610 Gabriel's second wife gave birth to a son, Michael, who was created a Baronet in 1627. He was a staunch republican in the reign of King Charles I and was a judge at the King's trial and one of those who signed the King's death warrant for his execution.

Gabriel, like his step-mother, Amy, left a generous legacy to St. Leonard's in his will dated 1628, comprising a tenement and land in Tooting Bec, the rents and profits from which were to be distributed among the poor of the parish at Easter and Christmas.

One of the properties in the tenement was a tavern, possibly called the Antelope, which is listed in *Taylor's Catalogue of Taverns* published in 1636. By 1761 this tavern had become the Bell public house.

This meant that for many years the church owned a pub in the parish. This caused some embarrassment in 1824 when the publican there, George Chaffy, was convicted, in the week before Christmas, of 'harbouring dissolute girls' in his establishment suggesting he was running a brothel, for which he was fined £5. The following year Chaffy was replaced as the licensee by Oliver Foster.

Next to the Bell once stood four small cottages. These were demolished by Henry Smith in 1822 and he built two larger cottages on the site at a cost of £600. These still stand today and Smith's initials and the date 1822 can be clearly seen on a small plaque between Nos 122 and 124 Upper Tooting Road.

In October 1887 the pub and Smith's cottages were sold to Miss Bell of Park Hill, Tooting, and the public house was demolished and replaced with a coffee house. The money from the sale was invested and the interest earned, together with the monies still payable under Amy Livesey's will, are administered today by the Streatham Charity Trustees.

The Bell Public House with Smith's cottages to the right

122-124 Upper Tooting Road, Smith's Cottages

ST. LEONARD'S CHURCH - STREATHAM

THE CRYPT

When the church was rebuilt in 1831 the architect advised that the foundations would have to be sunk between 10-15ft below ground level. It was therefore decided to create a crypt beneath the building in which to inter the intact remains of those persons buried under the floor of the old church and to sell off the surplus spaces to raise an estimated £1,000 to help defray the cost of the rebuilding work.

In addition, a large charnel pit was dug beneath the central aisle of the crypt in which bones unearthed in that part of the graveyard over which the new church was extended, and remains from beneath the old church floor, were deposited. The church accounts reveal that £2 1s 8d (£2.07) was spent in providing gin to the men who transferred the remains to the new vaults.

Entry to the charnel pit is obtained by raising the flagstones in the central aisle of the crypt, and large metal rings can be seen in the slabs to enable them to be lifted. This pit is known to be more than 21ft deep as during the First World War the then verger thrust drain rods to this length down through the bones and still did not touch the bottom of the pit!

Bones are still occasionally unearthed when road works are undertaken near the church. These have normally 'migrated' through the earth from the churchyard with the passing of time. These are now interred in a special coffin kept in the crypt for this purpose as, for safety reasons, the charnel pit is no longer in use.

In digging the crypt the workers reached the level of Roman Streatham and a number of coins from this period were discovered here. These, together with the discovery of a small Roman votive figure on the site of the Church of the English Martyrs, suggest this area may have been the site of a pagan shrine or place of worship in ancient times.

Ornate fittings and brass studs elaborately decorate some of the coffins in the crypt

Tomb of Thomas Forster, a chemist, who ran a rubber factory in Streatham which later became P B Cows and who was partner in a chemical works at Lonesome called Forster & Gregory

ST. LEONARD'S CHURCH - STREATHAM

The Thrale family vault in the crypt.

The creation of a crypt when the church was rebuilt in 1831 caused huge problems due to substandard workmanship by the builders.

Insufficient drainage resulted in the crypt being prone to flooding in periods of excessive wet weather and two ladies visiting the remains of their relatives in 1832 were almost drowned when they accidentally fell into a vault which was flooded with water!

The discovery that the brick piers supporting the church above were filled with earth rather than solid building material raised doubts they would be strong enough to carry the weight. It was feared that by breaking them open to rectify this more harm would be caused to the building than leaving them to stand as they were and this latter course of action was adopted.

However, the vaults in the north wall of the crypt were bricked up to provide additional support to the wall above. This led to the false belief that these were once secret tunnels leading to the Rectory and the Shrubbery.

Originally the crypt was kept open so that people could pay their respects to their departed loved ones as and when they wished and in the latter years of the 19th century, 'Black Tommy', a local tramp, lived here, and had letters addressed to him at St. Leonard's Crypt.

In the First and Second World Wars the crypt was used as an air raid shelter. During the Zeppelin Raids on Streatham the then Rector, Cannon Brooke Jackson, and his wife would wander among those sheltering here with sandwiches and mugs of hot tea and coffee and words of comfort.

Some of the tombs in the crypt have coffins stacked three or four high and many include small coffins containing young children who died in their infancy

Some of the coffins in the crypt have remained wonderfully preserved over the centuries and still show the splendour of the stud-work and beauty of the coffin plates, handles and artwork with which they were originally adorned.

Families could purchase a single space in the crypt for a loved one, or a full tomb or half tomb as required. Many of the vaults have metal gates with plaques identifying the space as belonging to a particular family.

Of special interest is vault No. 7, in the central aisle, in which are deposited the coffins from the Thrale family vault which previously was under the floor of the old church. Here you can see the massive coffin in which Henry Thrale was placed at the time of his death in 1781 (see page 44) and where he is destined to rest for eternity alongside his mother-in-law, Hester Salusbury; his wife having been buried with her second husband, Gabriel Piozzi, at Tremeirchoin Church in Wales.

After the fire in 1975 a spiral staircase was placed in an empty vault in the south-west corner of the south porch to allow internal access to the crypt. Prior to this the only entry was by way of the stairs leading down beneath the church to the left of the tower entrance to the building.

ST. LEONARD'S CHURCH - STREATHAM

REGISTERS

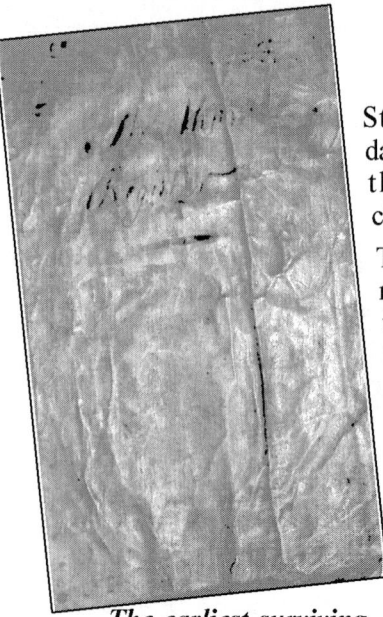

St. Leonard's registers date from 1538 and are the oldest surviving church records we have.

They list every baptism, marriage and burial undertaken at the church in almost 500 years with few exceptions, such as during the English Civil War when some of the entries are incomplete.

The last surviving burial register records interments up to 1862 the subsequent volume being lost. However, an additional 129 burials that took place between March 1862 and February 1944 are listed in other documents held in the parish archives.

In 1538 Thomas Cromwell ordered that every baptism, marriage and burial was to be registered weekly by the minister, with a Churchwarden acting as witness; these records were then to be deposited in the parish chest. These early records were often loose sheets of paper or parchment and very few originals survive from this time.

In 1598, Queen Elizabeth I ordered that all loose leaf registers were to be transcribed into parchment books from the commencement of her reign, hence the fact that a large number of surviving parish registers date from 1558.

St. Leonard's is fortunate that its registers were copied into a bound parchment register from 1538, giving us 20 years of earlier entries to refer to, and making Streatham one of only 656 parishes in Britain to have registers dating from this year.

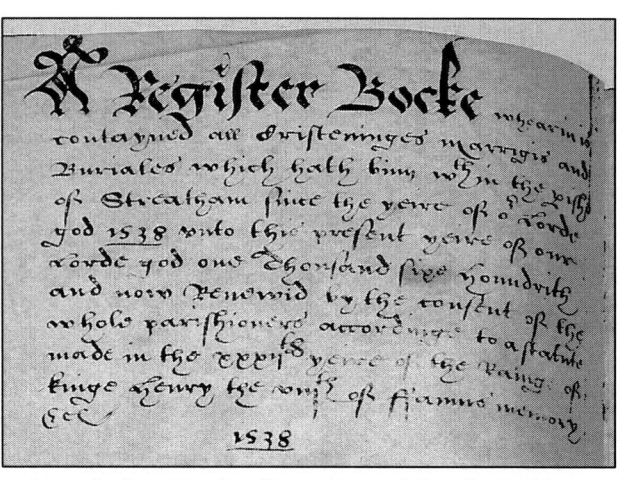

The earliest surviving Streatham Parish Register dating from 1538

Inscription on the first page of the Streatham Register for 1538 reads: 'A Register Book whear in is contayned all christenings marriages and burials which hath bin within the parish of Streatham since the year of our Lorde God 1538 unto the present year of our Lorde God one thousand six houndrith and now renewid by the consent of the whole parishioners according to a statute made in the XXXIIth year of the rainge of Kinge Henry the VIII of famus memory etc.'

The early entries comprise just one line, which details the date of the event and the names of the persons involved.

In 1644/5, it was ordered that in addition to the date of baptism, the date of birth should also be given, as also the names of both parents. Also the date of death was to be detailed as well as the day of burial.

The parish registers became more organised in 1711, when it was stipulated that proper printed books should be kept, with each page numbered and ruled.

In order to prevent a growing tide of clandestine marriages the Hardwicke Marriage Act was passed in 1754 which stipulated marriages could only be solemnised after banns had been published which were to be detailed in the back of the register or in

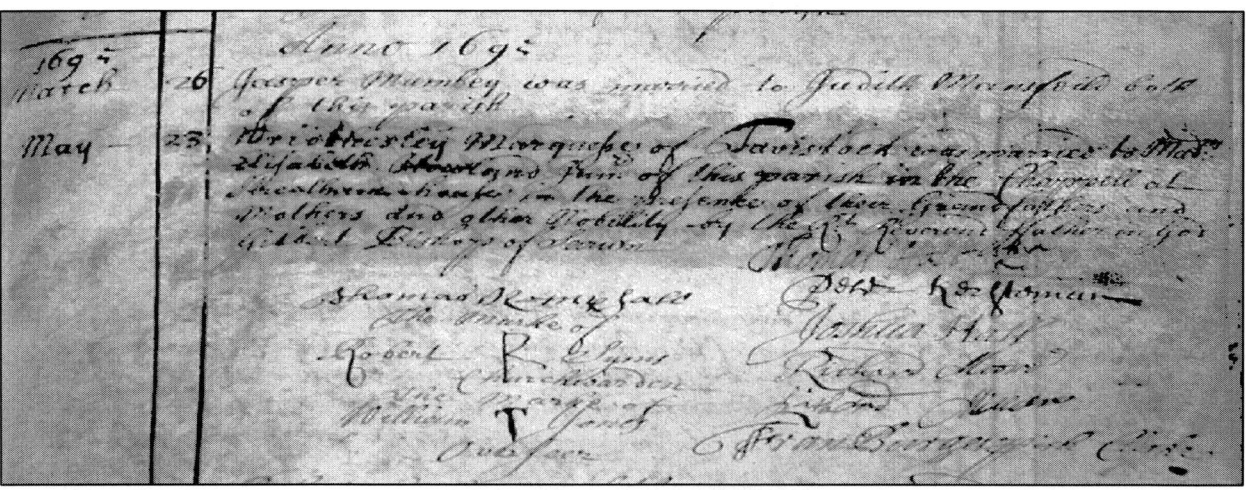

Entry in the parish register for 23rd May 1695 when the 14-year-old Marquess of Tavistock was married to 13-year-old Elizabeth Howland at Streatham Manor House (see pages 12 & 13)

a separate book. From this date the marriage register takes the form of printed pages in which the parish clerk filled in details of the marriage.

In 1812 printed registers were introduced for christenings and burials.

There were no further substantial changes until 1837 when the civil registration of births, marriages and deaths was introduced, causing another change to the layout of parish registers to conform to the new civil requirements.

The entries in the first St. Leonard's register are beautifully written, and suggests that a special scribe was employed to transcribe the entries from 1538 to 1598. The first entry is the baptism of Henry Holland on the 2nd January 1538.

There is also a lovely inscription in the register commencing in 1660 at the time of the restoration of the monarchy (see centre page). From this inscription it is obvious that the sympathies of the Rector lay with the Royalist cause.

The registers contain thousands of entries recorded over almost half a millennium with 14,084 baptisms taking place between 1538-1900; 4,630 marriages being solemnised between 1538-1899 and 10,682 burials being undertaken in the graveyard between 1538-1944.

A large number of strangers and travellers were baptised or buried at St. Leonard's as a consequence of two main high roads in Streatham and Tooting cutting through the Parish. On 11th August 1605 is the entry: 'Thomas Read son of John Read a stranger was baptised the same day whose mother fell into her trobell as she travelled by the highway', and on 20th April 1550 'Jane a pore wenche askinge her almes by the way' was buried.

There are also a number of instances where young children were abandoned by their parents, normally left at the church or on someone's doorstep, in the hope that a villager would take care of them as their parents were unable to do so. Normally the place where they were found was given to them as their surname such as a baptism on 20th July 1628 when 'Elizabeth Porch was baptised being left inside church' and on 29th April 1787 Mary Lyon was christened after being 'found at the White Lyon, parents unknown'.

During the 1690s there are a number of references to Dutch troopers, who presumably came to England with William III and were stationed in or near Streatham.

The old parish registers also contain a number of early entries relating to black people, the earliest of which dates from 1st September 1679 when 'Samuel - a black, moore servant to Mr William Prickman' was buried.

Several entries in the burial registers reveal how marriage ties were almost indissoluble even in death such as the burial on 3rd July 1679 of 'Joan Stratford, the wife of George Stratford, found dead in a well' and six months later on 4th December 1679 the burial of 'George Stratford he drowned himself in a well'.

Probably one of the most unusual entries relates to Elizabeth, the daughter of John Russell, who was baptised on 21st November 1669. This entry is marked 'The said Russell now living at Charing in Kent having sent for a copy of the register this 14th of September 1770', indicating that she was about to celebrate her 100th birthday later that year. She subsequently returned to Streatham where she died two years later aged 103. But on her death, the residents of Streatham were to get a rather unusual surprise as the entry in the burial register reveals: '14th April 1772 - RUSSELL NB This person was always known under the guise or habit of a woman, and answered to the name of Elizabeth, as registered in this parish November 21st 1669, but at death proved to be a man!'

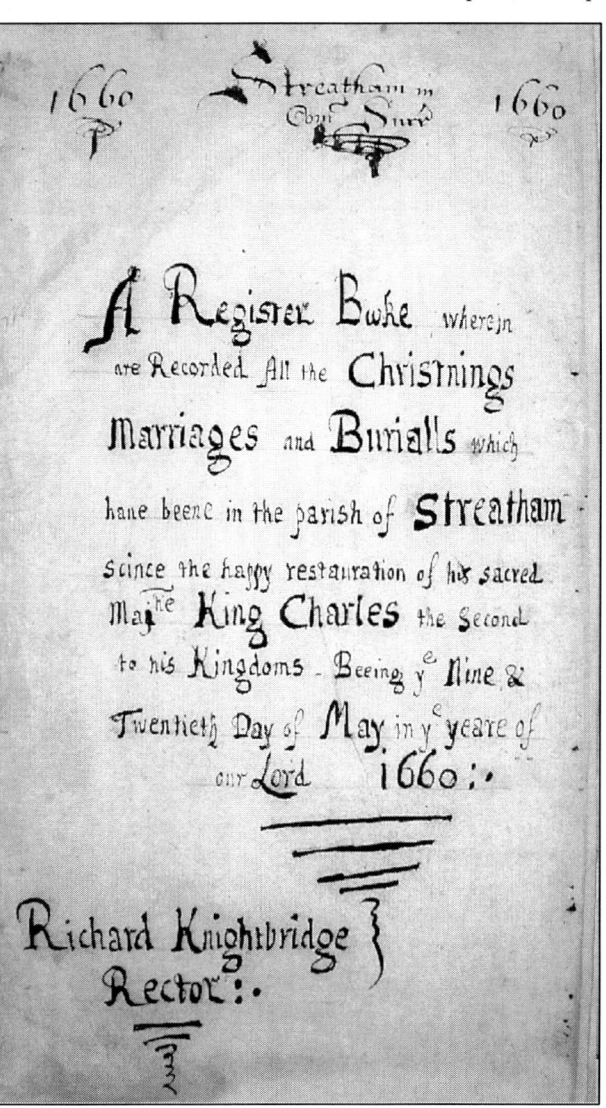

First page of the register commencing in 1660

CHURCH PLATE

The church plate mainly consists of items used in the communion service, comprising various chalices, patens, wafer boxes etc.

In 1547, the first year of the reign of Edward VI, an inventory of items of value held by the church was made. This included one chalice, two crosses of latten (a copper alloy such as brass or bronze), a latten basin, a cross cloth of silver, a little latten candlestick and three pewter cruets.

Four years later it was recorded that thieves broke into the church by smashing the east window and stole three latten candlesticks, two vestments and a 'sacryng bell' (a small bell which is rung when the Host is elevated at Mass).

This was not the only occasion when the church plate was stolen. In 1809 it was sent to Rundle and Bridge, silversmiths, in Ludgate Hill, London, to be repaired and cleaned. When this job had been completed it was packed into a box and sent back to Streatham on a horse cart. However, the cart driver left the box unattended in Bishopgate Street and it was stolen never to be seen again.

Among the existing church plate are a number of items given in memory of loved ones.

A large silver-gilt chalice, weighing 17 ounces, made in 1869 by S Smith was presented to St. Leonard's in memory of Eliza Haigh who died at Coventry House, a large mansion by Streatham Common, on 17th May 1876 aged 79. She was the widow of William Haigh of Furzedown House who died on 20th March 1861 aged 72. Both are buried in a large tomb in the graveyard with other members of the Haigh family. Furzedown House survives today and now forms part of Graveney School.

Another silver-gilt chalice, weighing 18 ounces, and a matching silver-gilt paten, made in Birmingham in 1869 by J Hardman & Co, were presented to the church in memory of Charles Payne who died on 20th June 1869.

A silver wafer box is in memory of Eric Charles Duncan Hollis who died of pneumonia at the tragically young age of 17 on 2nd January 1932. He was a server at St. Leonard's church and the box was presented by his family and fellow servers at the church.

Also among the plate is an 8 inch diameter silver-gilt paten made in 1854 by R. Garrard. This plate has six arcs with points at the angles with a sacred monogram in the middle. There is an engraved floriated ornament round the rim of the plate. It is possible this item may have been presented to the church by its maker, Robert Garrard, the Crown Jeweller, who lived in the parish at Woodfield Lodge, by Tooting Bec Common, was churchwarden of St. Leonard's for five years from 1848-1852 and is buried in the graveyard (see page 61).

Silver chalis presented in memory of Eliza Haigh in 1876

Silver chalis presented in memory of Charles Payne in 1869

Wafer box presented in 1932 in memory of Eric Charles Duncan Hollis

Silver paten made in 1854 by Robert Garrard who was churchwarden at St. Leonard's from 1848-52

ST. LEONARD'S CHURCH - STREATHAM

THE GRAVEYARD

In death all men and women are equal, and no where is this more evident in Streatham than in our ancient parish graveyard. Clerk to the Privy Council, the Crown Jeweller, and Lords of the Manor lie alongside the village rat catcher, the local beer seller and the unmarked graves of countless paupers.

Each grave has its own tale to tell. Sometimes this is carved clearly for all to see, like the grave of Mary Ann West West who died on 16th June 1865 'from falling from the cliff above Llandudno'. For others their secrets lie buried in the old parish records. However, for the majority, what they did, where they lived and how they died are now lost in the antiquity of time.

282 graves have been identified in the churchyard. The earliest is that of Humphrey Townsend who died on 1st November 1708 aged 17, the latest is that of the Montefiore family which was the last grave to be opened, when in February 1944 90 year old Edith Montefiore was interred there.

We know from the parish registers, and other sources, that between 1538 and 1944 10,682 people were buried in the small graveyard surrounding St. Leonard's church.

Gravestone for Sarah Hale, who died in 1768, aged 27

Obviously over the years the graves of many of these people have disappeared as the plots have been reused and their memorials have disintegrated. Like that of Richard Clarke laid to rest on 24th September 1690 after being 'killed coming down from Croydon Fair'.

In February 1814 there was an attempt to steal a freshly interred body from the graveyard when Thomas Watts was caught digging up one of the graves. He, and his two accomplices, Thomas Butler and William Lane, all residents of Southwark, each received a sentence of 3 months imprisonment for attempted body snatching.

In 1831, when the church was rebuilt, tall railings were placed around the graveyard at a cost of £480 to deter the grave robbers. These railings were removed in 1942 when they were sold as 13½ tons of scrap metal to help the war effort for which the church received £18 13s 9d (£18.69) in compensation.

By Act of Parliament, in 1875, St. Leonard's graveyard was closed for new burials, and only those who had already purchased a grave were allowed to be interred there. However, a special dispensation was given so that the church Sexton, Henry Daniels, who had served the church for

The Montefiore family tomb, the last grave to be opened in the churchyard, in 1944

Photograph taken in the early 1900s showing the tall railings which were placed around the graveyard in 1831 to deter grave robbers

The three Grade II listed monuments in the churchyard. From left to right, coade stone tomb of Joseph Hay, who died 6th August 1805, aged 50; tomb of George Abell, who died 25th August 1826, aged 36 and tomb of Lt. Col. William Boyce of the 16th Queen's Light Dragoons who died 17th December 1808, aged 66

almost 50 years until his death in 1878, at the age of 76, could be buried in the churchyard he had looked after for almost half a century.

On the 12th August 1891 the Streatham Burial Board purchased just over 23 acres of land forming part of Springfield Farm in Garratt Lane, Tooting, for £17,684 5s. 0d. (£17,684.25) for use as a parish graveyard and this was opened as Streatham Cemetery in 1893. This continues in use today and is now administered by Lambeth Council.

Large areas of St. Leonard's graveyard suffered extensive damage during the Second World War. On the 29th October 1940 a bomb fell in the north-eastern corner of the churchyard demolishing many graves as well as destroying the stained glass windows at the eastern end of the church. Then on the 10th November 1940 another bomb fell at the junction of Streatham High Road and Gleneldon Road, the blast from which destroyed many of the graves on the southern side of the church.

In 1958 a Garden of Remembrance was established in the north-eastern corner of the churchyard where ashes of the cremated can be buried. The Garden was laid out on the site of the grave of Robert Gibson Esq of Upper Tooting who died in 1821, and whose grave was one of those destroyed by bombing in the war.

As from 10th May 1971 the churchyard was closed for all further burials, or re-opening of graves, and now only the interment of ashes is permitted.

During 1999 a major restoration of the graveyard took place involving the re-laying of paths, the installation of external lighting to the church and graveyard paths, the erection of railings and gates around the perimeter of the graveyard and the general landscaping of the area.

Today the churchyard is a natural oasis with 109 different plants, shrubs and trees identified by members of the South London Botanical Institute (SLBI).

It also provides an important sanctuary for insects and wildlife with 15 different species of bird identified including Greenfinches, Magpies, Ring-necked parakeets, Robins and Wrens. Sparrows are now rarely seen as a Sparrow Hawk has recently been seen circling the church.

Three of the tombs in the graveyard are Grade II listed monuments. These are the tombs of George Abell, Lt. Col. William Boyce and Joseph Hay.

Notable among those buried in our churchyard are:

WILLIAM MATTHEW COULTHURST died 10th February 1877 aged 84 years. Senior partner in Coutts Bank and a generous local philanthropist. Lived at Streatham Lodge, South Streatham, from 1836-1877. A generous supporter of the work of Immanuel Church, Streatham Common, of which he was churchwarden for over 20 years.

BERIAH DREW died 17th August 1878 aged 90. In 1836 he became Lord of the Manor of Leigham and one of the largest landowners in the parish. He developed Leigham Court Road at Streatham Hill and Drewstead Road is named in his honour.

WILLIAM DYCE died 11th February 1864 aged 58. Artist and member of the Royal Academy and designer of the florin (10p) coin. His murals can be seen today in the Houses of Parliament, Buckingham Palace and Osborne House on the Isle of Wight. He was churchwarden of St. Leonard's from 1862 until his death and designed the chancel extension to the Church. In appreciation of his work for the church the parishioners erected the Dyce Fountain in his honour, which he designed himself, and now stands in the centre of Streatham Green.

ROBERT GARRARD died 16th September 1881 aged 88. Crown jeweller and maker of many sporting trophies including the Ascot Cup and the America's Cup, the oldest international sporting trophy in the world. He was churchwarden of St. Leonard's church. Garrads Road, along the western end of Tooting Bec Common, is named in his honour, although the third R in his surname was omitted in error when naming the road! (see page 58)

SIR ARTHUR HELPS KCB died 7th March 1875 aged 61. Clerk to the Privy Council and an important confident to Queen Victoria after the death of Prince Albert. In 1862 the Queen asked him to revise Prince Albert's speeches for publication and he also edited the Queen's *Highland Journal* which was published in 1868.

ST. LEONARD'S CHURCH - STREATHAM

JOHN HONE died 20th December 1812 aged 86. He was the Beadle of Streatham and was responsible for keeping the poor of the parish in order. He also acted as a church messenger and parish policeman. He wore an impressive yellow coat and three cornered hat, both of which were trimmed with gold braid. He escorted vagrants and beggars either out of the Parish or to the workhouse which stood on Tooting Bec Common. When he died he was succeeded by his son, John junior, whose wife, Mary, had a reputation for being a miser. On her death on 20th May 1836, aged 74, it was discovered her mattress was stuffed with money which she had hoarded during her lifetime.

SARAH THOMPSON STREET, wife of William Street, died 25th July 1825 aged 66. In the 1780s William was vestry clerk at St. Leonard's church and as well as being responsible for the parish records he also made the parish Beadle's coat which in 1786 he dressed with gold braid and lace for £6. He charged separately for pockets which cost an extra 7s (35p) each. Part of his duties also involved the winding of the church clock in the tower for which he was paid £2 12s 6d (£2.62½) per annum in 1785. He developed a novel way of keeping the unemployed in Streatham occupied, for the church ledgers show he was paid 5s (25p) in 1791 for employing two men 'for digging holes and filling them up again'.

GEORGE PRATT died 14th March 1890 aged 62. A successful retailer and developer of the Bedford Park Estate. He came to Streatham in 1839 as a 13-year-old apprentice in William Reynold's small drapery shop situated opposite Streatham Green. When Reynolds retired George bought the business and built it up to become the largest shop in the town - Pratts Department Store. This store eventually traded as part of the John Lewis Partnership until it closed in 1990.

George Pratt standing at the entrance to his store in Streatham High Road in the 1870s

GEORGE FRANCIS TROLLOPE died 13th March 1895 aged 78. Builder whose family business later became Trollope & Colls, a major construction and building company. He lived at Elmfield, Leigham Court Road, and his company built many of the large houses in this road. It also erected St. Peter's church there in 1870, after which George and his family played an active part in the church they had built.

St. Pater's Church

THE GLEBE

Standing to the north of St. Leonard's church in Tooting Bec Gardens is an area of open ground known as the Glebe.

The word Glebe comes from the Latin g'ba which means soil. Glebe land existed in most parishes and was set aside for the upkeep and maintenance of the parish priest. The land was originally used for growing crops or grazing livestock to provide sustenance for the priest.

The Glebe in Streatham originally covered a large area and, in 1769, the then Rector, James Tattersall, built a large and imposing house on part of the land which later became known as the Shrubbery (see page 27). This was a very ornate building with Adam fireplaces and intricate plaster work on the walls and ceilings.

The Revd. Tattersall filled his garden with a large number of statues and garden ornaments. On Sunday 23rd August 1778 Fanny Burney and Hester Thrale visited the Shrubbery, and Fanny recalled the occasion in her diary:

'After dinner I had a delightful stroll with Mrs. Thrale, and she gave me a list of all her good neighbours in the town of Streatham and she said she was determined to take me to see Mr Tattersall, the clergyman, who was a character and could not but be diverted with, for he had so furious and so absurd a rage for building that in his garden he had as many temples and summer houses and statues as in the gardens of Stow, though he had so little room for them that they all seemed tumbling one upon another.'

In 1894 the Shrubbery was sold to the Church School Society which established a Girls' School and kindergarten there known as the Streatham High School for Girls and later as Streatham College for Girls. Such was the 'quality' of the young ladies who attended the school that local children referred to the building as 'The Snobbery'.

This house was demolished in 1933 and a parade of shops with flats above, known as Central Parade, 254 Streatham High Road, designed by Dixon and Braddock ARIBA was erected on the site in 1934.

By the mid 1800s the Glebe had shrunk in size to around 1½ acres and had been absorbed into the garden of the Rectory, which occupied the site where St. Leonard's Church Hall now stands.

In the late 1970s it was decided to build sheltered housing on part of the Glebe and in 1977 Lambeth Council erected 34 flats on the Prentis Road side of the Glebe, together with special accommodation for disabled people, and communal facilities including a club room.

The estate was designed by the architects, Stillman and Eastwick-Field, and as many as possible of the surviving trees and natural features were incorporated into the development.

Since the development was opened in 1978 it has won numerous awards, including commendations in the Housing Design Awards, The Civic Trust Awards and from the Royal Institute of British Architects.

The Glebe now provides the church with a small outdoor area for various parish events. In 1981 a number of trees were planted here including, Silver Birch, Mountain Ash, Yew, and Holm Oak.

At Easter 2013 a donkey led a procession from Streatham Green to St. Leonard's church for the Palm Sunday service and a new, large wooden cross was erected on the Glebe to replace the previous cross, the base of which had rotted away.

The wooden cross on the Glebe.

The Shrubbery

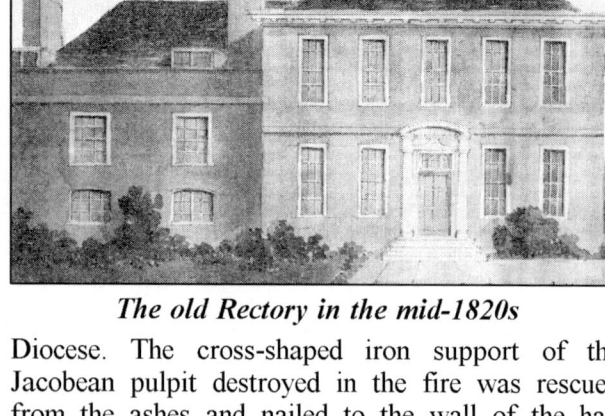

CHURCH HALL & RECTORY

The old Rectory in the mid-1820s

The hall was built in 1908 on the site of the old Rectory as a memorial to Canon John Nicholl, Streatham's longest serving Rector.

The old Rectory dated back to at least 1535 when Thomas Martyn was vicar. It was enlarged and altered several times over the centuries.

It is claimed that in the study of the Rectory Lord John Russell worked on preliminary drafts of the great Reform Bill of 1832. Seeking a quiet retreat where he could work undisturbed he stayed there during the period when his brother, the Revd. Wriothesley Russell, was Rector of Streatham between 1830-35.

When the Rectory was demolished in 1907, to make way for the new church hall, a new Rectory was built on the Glebe facing Tooting Bec Gardens. This was pulled down in 1972 and a house was purchased in Becmead Avenue which currently serves as the Rectory.

After the fire which destroyed the church in May 1975 the Hall was used for services and became known as the Hall Church. This was fitted out through the generosity of other churches in the Diocese. The cross-shaped iron support of the Jacobean pulpit destroyed in the fire was rescued from the ashes and nailed to the wall of the hall above the altar to serve as a crucifix. Services were held here until the church had been rebuilt and was brought back into full-time use in May 1977.

The Hall Church in 1975

The church hall is currently the headquarters of the Spires Centre which was founded in 1990 to provide meals for homeless people over Christmas.

It now operates a day centre providing essential services to disadvantaged members of the local community.

In 2014 over 800 disadvantaged men and women were assisted by the Spires Centre and its Rough Sleepers Space, which provides food, showers and clothing in addition to an advisory service, supported 518 homeless people.

Spires also supported 92 women involved in street-based sex work in 2014 through its Streetlink project.

The Church Hall, Rectory and St. Leonard's Church c1910